Probability
DeMYSTiFieD®

DeMYSTiFieD® Series

Accounting Demystified

Advanced Calculus Demystified

Advanced Physics Demystified

Advanced Statistics Demystified

Algebra Demystified

Alternative Energy Demystified

Anatomy Demystified

Astronomy Demystified

Audio Demystified

Biology Demystified

Biophysics Demystified

Biotechnology Demystified

Business Calculus Demystified

Business Math Demystified

Business Statistics Demystified

C++ Demystified

Calculus Demystified

Chemistry Demystified

Chinese Demystified

Circuit Analysis Demystified

College Algebra Demystified

Corporate Finance Demystified

Data Structures Demystified

Databases Demystified

Differential Equations Demystified

Digital Electronics Demystified

Earth Science Demystified

Electricity Demystified

Electronics Demystified

Engineering Statistics Demystified

Environmental Science Demystified

Ethics Demystified

Everyday Math Demystified

Financial Planning Demystified

Forensics Demystified

French Demystified

Genetics Demystified

Geometry Demystified

German Demystified

Home Networking Demystified

Investing Demystified

Italian Demystified

Japanese Demystified

Java Demystified

JavaScript Demystified

Lean Six Sigma Demystified

Linear Algebra Demystified

Logic Demystified

Macroeconomics Demystified

Management Accounting Demystified

Math Proofs Demystified

Math Word Problems Demystified

MATLAB® Demystified

Medical Billing and Coding Demystified

Medical Terminology Demystified

Meteorology Demystified

Microbiology Demystified

Microeconomics Demystified

Minitab Demystified

Nanotechnology Demystified

Nurse Management Demystified

OOP Demystified

Options Demystified

Organic Chemistry Demystified

Personal Computing Demystified

Pharmacology Demystified

Philosophy Demystified

Physics Demystified

Physiology Demystified

Pre-Algebra Demystified

Precalculus Demystified

Probability Demystified

Project Management Demystified

Psychology Demystified

Quality Management Demystified

Quantum Mechanics Demystified

Real Estate Math Demystified

Relativity Demystified

Robotics Demystified

Signals and Systems Demystified

Six Sigma Demystified

Spanish Demystified

Statics and Dynamics Demystified

Statistical Process Control Demystified

Statistics Demystified

Technical Analysis Demystified

Technical Math Demystified

Trigonometry Demystified

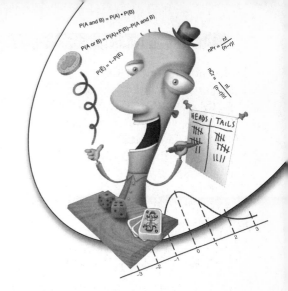

$$P(A \text{ and } B) = P(A) \cdot P(B)$$

$$P(A \text{ or } B) = P(A) + P(B) - P(A \text{ and } B)$$

$$P(\bar{E}) = 1 - P(E)$$

$$nPr = \frac{n!}{(n-r)!}$$

$$nCr = \frac{n!}{(n-r)!r!}$$

HEADS | TAILS

Probability
DeMYSTiFieD®

Allan G. Bluman

Second Edition

Mc
Graw
Hill

New York   Chicago   San Francisco   Lisbon   London   Madrid   Mexico City
Milan   New Delhi   San Juan   Seoul   Singapore   Sydney   Toronto

McGraw-Hill books are available at special quantity discounts to use as premiums and sales promotions, or for use in corporate training programs. To contact a representative, please e-mail us at bulksales@mcgraw-hill.com.

Probability DeMYSTiFieD®, Second Edition

ISBN 978-0-07-178097-1
MHID 0-07-178097-1

Sponsoring Editor
Judy Bass

Editorial Supervisor
Stephen M. Smith

Production Supervisor
Pamela A. Pelton

Acquisitions Coordinator
Bridget L. Thoreson

Project Manager
Kritika Kaul,
Cenveo Publisher Services

Copy Editor
Medha Joshi,
Cenveo Publisher Services

Proofreader
Ash Janmeja,
Cenveo Publisher Services

Cover Illustration
Lance Lekander

Art Director, Cover
Jeff Weeks

Composition
Cenveo Publisher Services

To all of my teachers, whose examples instilled
in me my love of mathematics and teaching.

About the Author

Allan G. Bluman taught mathematics and statistics in high school, college, and graduate school for over 35 years. He received his doctorate from the University of Pittsburgh. He has written three textbooks for McGraw-Hill, and he is also the author of three other books in the *Demystified*® series: *Pre-Algebra Demystified*, *Math Word Problems Demystified*, and *Business Math Demystified*. Dr. Bluman is the recipient of an "Apple for the Teacher" award for bringing excellence to the learning environment, and he has received two "Most Successful Revision of a Textbook" awards from McGraw-Hill. His biographical record appears in *Who's Who in American Education*, Fifth Edition.

Contents

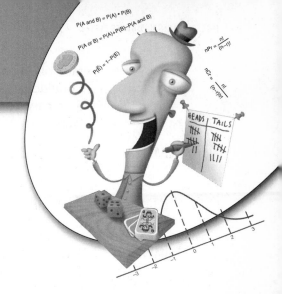

Preface

"The probable is what usually happens."—Aristotle

Probability can be called the mathematics of chance. The theory of probability is unusual in the sense that we cannot predict with certainty the individual outcome of a chance process such as flipping a coin or rolling a die (singular for dice), but we can assign a number that corresponds to the probability of getting a particular outcome. For example, the probability of getting a head when a coin is tossed is $\frac{1}{2}$ and the probability of getting a two when a single fair die is rolled is $\frac{1}{6}$.

We can also predict with a certain amount of accuracy that when a coin is tossed a large number of times, the ratio of the number of heads to the total number of times the coin is tossed will be close to $\frac{1}{2}$.

Probability theory is, of course, used in gambling. Actually, mathematicians began studying probability as a means to answer questions about gambling games. Besides gambling, probability theory is used in many other areas such as insurance, investing, weather forecasting, genetics, and medicine, and in everyday life.

What Is This Book About?

First let me tell you what this book is **not** about.

- This book is **not** a rigorous theoretical deductive mathematical approach to the concepts of probability.
- This book is **not** a book on how to gamble.

And most important,

- This book is **not** a book on how to win at gambling!

This book presents the basic concepts of probability in a simple, straightforward, easy-to-understand way. It does require, however, knowledge of arithmetic (fractions, decimals, and percents) and knowledge of basic algebra (formulas, exponents, order of operations, etc.). If you need a review of these concepts, you can consult another of my books in this series entitled *Pre-Algebra Demystified*.

This book can be used to gain knowledge of the basic concepts of probability theory, either as a self-study guide or as a supplementary textbook for those who are taking a course in probability or a course in statistics that has a section on probability.

The basic concepts of probability are explained in the first two chapters. Then the addition and multiplication rules are explained. Following that, the concepts of odds and expectation are explained. The counting rules are explained in Chapter 6, and they are needed for the binomial and other probability distributions found in Chapters 7 and 8. The relationship between probability and the normal distribution is presented in Chapter 9. Finally, a recent development, the Monte Carlo method of simulation, is explained in Chapter 10. Chapter 11 explains how probability can be used in game theory, and Chapter 12 explains how probability is used in actuarial science. Special material on Bayes' theorem is presented in the Appendix because this concept is somewhat more difficult than the other concepts presented in this book.

Near the end of each chapter is what is called a "Probability Sidelight." These sections cover some of the historical aspects of the development of probability theory or some commentary on how probability theory is used in gambling and everyday life.

I have spent my entire career teaching mathematics at a level that most students can understand and appreciate. I have written this book with the same objective in mind. Mathematical precision, in some cases, has been sacrificed in the interest of presenting probability theory in a simplified way.

Good luck!

Curriculum Guide

The *Demystified®* books are closely linked to the standard high school and college curricula, so the Curriculum Guide found on the inside back cover is provided to give you a clear path to meet your mathematical goals. What many students do not know is that mathematics is a hierarchical subject. What this means is that before you can be successful in algebra, you need to know basic arithmetic, since the concepts of arithmetic (pre-algebra) are used in algebra. Before you can be successful in trigonometry, you need to have a basic understanding of algebra and geometry, since trigonometry uses concepts from these two courses, etc. You can use the Guide in your mathematical studies to learn which courses are necessary before you take the next one.

How to Use This Book

As you know, in order to build a tall building, you need to start with a strong foundation. It is also true in mastering mathematics that you need to start with a strong foundation. This book presents the basic topics in probability in a logical, easy-to-read format. This book can be used as an independent study course or as a supplement to a probability course.

To learn mathematics, you must know the vocabulary, understand the rules and procedures, and be able to apply these rules and procedures to mathematical problems in order to solve them. This book is written in a style that will help you with learning. Important terms have been boldfaced and important rules and procedures have been italicized. Each section has several worked-out examples showing you how to use the rules and procedures. Each section also contains several practice problems for you to work out to see if you understand the concepts. The correct answers are printed immediately after the problems so you can see if you have solved them correctly. At the end of each chapter is a multiple-choice quiz. If you answer most of the problems correctly, you can move on to the next chapter. If not, please repeat the chapter. Make sure you do not look at the answer before you have attempted to solve the problem.

Even if you know some or all of the material in a chapter, it is best to work through the chapter in order to review the material. The little extra effort will be a great help when you encounter more difficult material later. After you complete the entire book, you can take the final exam and determine your level of competence. You should use a calculator to help you with the computations and save time.

For this second edition, most of the examples and exercises have been changed. In addition, explanations called "Still Struggling?" have been added to help you gain a better understanding of the concepts.

Best wishes on your success!

Allan G. Bluman

Acknowledgments

I would like to thank my wife, Betty Claire, for helping me with the preparation of this book and my editor, Judy Bass, for her assistance in its publication. I would also like to thank Eugene Mastroianni for his error checking and helpful suggestions.

Probability
DeMYSTiFieD®

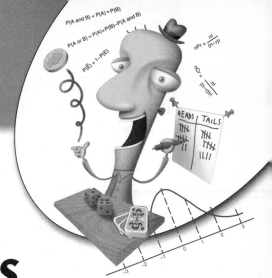

chapter 1

Basic Concepts

Probability can be defined as the mathematics of chance. Most people are familiar with some aspects of probability by observing or playing gambling games such as lotteries, slot machines, blackjack, or roulette. However, probability theory is used in many other areas such as business, insurance, weather forecasting, and in everyday life.

In this chapter, you will learn about the basic concepts of probability using various devices such as coins, cards, and dice. These devices are not used as examples in order to make you an astute gambler, but they are used because they will help you understand the concepts of probability.

CHAPTER OBJECTIVES

In this chapter, you will learn

- The basic concepts of probability including probability experiments, sample spaces, simple and compound events, and equally likely events
- How to find the probability of an event using the classical probability formula
- How to find the probability of an event using a frequency distribution
- The range of probability values
- The basic probability rules
- How to find the probability of a complement of an event
- The law of large numbers
- The concept of subjective probability

Probability Experiments

Chance processes, such as flipping a coin, rolling a die (singular for dice), or drawing a card at random from a well-shuffled deck are called *probability experiments*. A **probability experiment** is a chance process that leads to well-defined outcomes or results. For example, tossing a coin can be considered a probability experiment since there are two well-defined outcomes—heads and tails.

An **outcome** of a probability experiment is the result of a single trial of a probability experiment. A **trial** means flipping a coin once or drawing a single card from a deck. A trial could also mean rolling two dice at once, tossing three coins at once, or drawing five cards from a deck at once. A single trial of a probability experiment means to perform the experiment one time.

The set of all outcomes of a probability experiment is called a **sample space**. Some sample spaces for various probability experiments are shown here.

Experiment	Sample Space
Toss one coin	H, T*
Roll a die	1, 2, 3, 4, 5, 6
Toss two coins	HH, HT, TH, TT
*H = heads; T = tails	

Notice that when two coins are tossed, there are four outcomes, not three. Consider tossing a nickel and a dime at the same time. Both coins could fall heads up. Both coins could fall tails up. The nickel could fall heads up and the dime could fall tails up, or the nickel could fall tails up and the dime could fall heads up. The situation is the same even if the coins are indistinguishable.

It should be mentioned that each outcome of a probability experiment occurs at **random**. This means you cannot predict with certainty which outcome will occur when the experiment is conducted. Also, each outcome of the experiment is **equally likely** unless otherwise stated. That means that each outcome has the same probability of occurring.

When finding probabilities, it is often necessary to consider several outcomes of the experiment. For example, when a single die is rolled, you may want to consider obtaining an even number; that is, a 2, 4, or 6. This is called an event. An **event** then consists of one or more outcomes of a probability experiment.

NOTE   *It is sometimes necessary to consider an event which has no outcomes. This will be explained later.*

An event with one outcome is called a **simple event**. For example, a die is rolled and the event of getting a 4 is a simple event since there is only one way to get a 4. When an event consists of two or more outcomes, it is called a **compound event**. For example, if a die is rolled and the event is getting an odd number, the event is a compound event since there are three ways to get an odd number, namely, 1, 3, or 5.

Simple and compound events should not be confused with the number of times the experiment is repeated. For instance, if two coins are tossed, the event of getting two heads is a simple event since there is only one way to get two heads, whereas the event of getting a head and a tail in either order is a compound event since it consists of two outcomes, namely, head, tail and tail, head.

EXAMPLE

A single die is rolled. List the outcomes in each event:

   a. Getting an even number

   b. Getting a number greater than 3

   c. Getting less than 1

SOLUTION

   a. The event contains the outcomes of 2, 4, and 6

   b. The event contains the outcomes of 4, 5, and 6

   c. When you roll a die, you cannot get a number less than 1; hence, the event contains no outcomes

 Still Struggling

Remember, when finding the outcomes in the sample space, be sure to include all the outcomes. For example, when two coins are tossed, there are four outcomes, not three, since there are two ways to get a head and a tail: HT and TH.

Classical Probability

Sample spaces are used in **classical probability** to determine the numerical probability that an event will occur. The formula for determining the probability of an event E is

$$P(E) = \frac{\text{number of outcomes contained in the event } E}{\text{total number of outcomes in the sample space}}$$

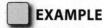 **EXAMPLE**

Two coins are tossed; find the probability that both coins land tails up.

 **SOLUTION**

The sample space for tossing two coins is HH, HT, TH, and TT. Since there are four events in the sample space, and only one way to get two tails (TT), the answer is

$$P(\text{TT}) = \frac{1}{4}$$

EXAMPLE

A die is tossed; find the probability of each event:
 a. Getting a 5
 b. Getting an odd number
 c. Getting a number less than 4

SOLUTION

The sample space is 1, 2, 3, 4, 5, 6, so there are six outcomes in the sample space.

   a. $P(5) = \frac{1}{6}$, since there is only 1 way to obtain a 5
   b. $P(\text{odd number}) = \frac{3}{6} = \frac{1}{2}$, since there are 3 ways to get an odd number, 1, 3, or 5
   c. $P(\text{number less than 4}) = \frac{3}{6} = \frac{1}{2}$, since there are 3 numbers in the sample space less than 4

**EXAMPLE**

A dish contains six red jellybeans, four yellow jellybeans, two black jelly-beans, and three pink jellybeans. If a jellybean is selected at random, find the probability that it is

  a. A pink jellybean

  b. A yellow or black jellybean

  c. Not a red jellybean

  d. A blue jellybean

SOLUTION

There are $6 + 4 + 2 + 3 = 15$ outcomes in the sample space.

  a. $P(\text{pink}) = \frac{3}{15} = \frac{1}{5}$

  b. $P(\text{yellow or black}) = \frac{4+2}{15} = \frac{6}{15} = \frac{2}{5}$

  c. $P(\text{not red}) = P(\text{yellow or black or pink}) = \frac{4+2+3}{15} = \frac{9}{15} = \frac{3}{5}$

  d. $P(\text{blue}) = \frac{0}{20} = 0$, since there are no blue jellybeans

Probabilities can be expressed as reduced fractions, decimals, or percents. For example, if a coin is tossed, the probability of getting heads up is $\frac{1}{2}$ or 0.5 or 50%.

NOTE *Some mathematicians feel that probabilities should be expressed only as fractions or decimals. However, probabilities are often given as percents in every-day life. For example, one often hears, "There is a 50% chance that it will rain tomorrow."*

Probability problems use a certain language. For example, suppose a die is tossed. An event that is specified as "getting at least a 3" means getting a 3, 4, 5, or 6. An event that is specified as "getting at most a 3" means getting a 1, 2, or 3.

Probability Rules

There are certain rules that apply to classical probability theory. They are pre-sented next.

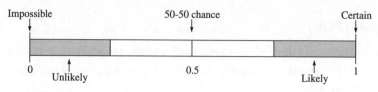

FIGURE 1-1

Rule 1: *The probability of any event will always be a number from 0 to 1.*

This can be denoted mathematically as $0 \leq P(E) \leq 1$. What this means is that all answers to probability problems will be a number ranging from 0 to 1. Probabilities cannot be negative nor can they be greater than one.

Also, when the probability of an event is close to zero, the occurrence of the event is relatively unlikely. For example, if the chances that you will win a certain lottery are 0.001 or 1 in 1000, you probably won't win, unless of course, you are very "lucky." When the probability of an event is 0.5 or $\frac{1}{2}$, there is a 50-50 chance that the event will happen—the same probability of getting one of the two outcomes when flipping a coin. When the probability of an event is close to one, the event is almost certain to occur. For example, if the chance of it snowing tomorrow is 90%, more than likely you'll see some snow. See Figure 1-1.

Rule 2: *When an event cannot occur, the probability will be 0.*

EXAMPLE _____

A die is rolled; find the probability of getting a 7.

SOLUTION _____

Since the sample space is 1, 2, 3, 4, 5, and 6, and there is no way to get a 7, $P(7) = 0$. The event in this case has no outcomes which are contained in the sample space.

Rule 3: *When an event is certain to occur, the probability is 1.*

EXAMPLE _____

A die is rolled; find the probability of getting a number less than 7.

SOLUTION

Since all outcomes in the sample space are less than 7, the probability is $\frac{6}{6} = 1$.

Rule 4: *The sum of the probabilities of all of the outcomes in the sample space is 1.*

Refer to the sample space for tossing two coins (HH, HT, TH, TT): Each outcome has a probability of $\frac{1}{4}$, and the sum of the probabilities of all of the outcomes is $\frac{1}{4} + \frac{1}{4} + \frac{1}{4} + \frac{1}{4} = \frac{4}{4} = 1$.

Rule 5: *The probability that an event will not occur is equal to 1 minus the probability that the event will occur.*

For example, when a die is rolled, the sample space is 1, 2, 3, 4, 5, 6. Now consider the event E of getting a number less than 3. This event consists of the outcomes 1 and 2. The probability of event E is $P(E) = \frac{2}{6} = \frac{1}{3}$. The outcomes in which E will not occur are 3, 4, 5, and 6, so the probability that event E will not occur is $\frac{4}{6} = \frac{2}{3}$. The answer to $P(\text{not } E) = 1 - P(E) = 1 - \frac{1}{3} = \frac{2}{3}$.

If an event E consists of certain outcomes, then event $\bar{E}$ (E bar) is called the **complement** of event E and consists of the outcomes in the sample space which are not outcomes of event E. In the previous situation, the outcomes in E are 1 and 2. Therefore, the outcomes in $\bar{E}$ are 3, 4, 5, and 6. Now Rule 5 can be stated mathematically as $\bar{E} = 1 - P(E)$.

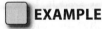 **EXAMPLE**

If the chance of rain is 0.40 (40%), find the probability that it won't rain.

 **SOLUTION**

Since $P(E) = 0.40$ and $P(\bar{E}) = 1 - P(E)$, then the probability that it won't rain is $1 - 0.40 = 0.60$ or 60%. Hence the probability that it won't rain is 60%.

 ## Still Struggling

Even though an event cannot occur, we still can assign a probability value mathematically. It is 0.

PRACTICE

1. A single die is rolled. Find each probability:

   a. The number shown on the face is a 4.

   b. The number shown on the face is less than 3.

   c. The number shown on the face is less than 1.

   d. The number shown on the face is divisible by 3.

2. A box contains a $1 bill, a $2 bill, a $5 bill, a $10 bill, and a $20 bill. A person selects a bill at random. Find each probability:

   a. The bill selected is a $2 bill.

   b. The denomination of the bill selected is more than $5.

   c. The bill selected is a $100 bill.

   d. The bill selected is of an even denomination.

   e. The denomination of the bill is divisible by 10.

3. Two coins are tossed. Find each probability:

   a. Getting two heads.

   b. Getting at least one tail.

   c. Getting two tails.

4. The cards A♥, 2♦, 3♣, 4♥, 5♠, and 6♣ are shuffled and dealt face down on a table. (Hearts and diamonds are red; clubs and spades are black.) If a person selects one card at random, find the probability that the card is

   a. The ace of hearts.

   b. A black card.

   c. A spade.

5. A spinner for a child's game has the numbers 1 through 5 evenly spaced. If a child spins, find each probability:

   a. The number is divisible by 2.

   b. The number is greater than 6.

   c. The number is an odd number.

6. A letter is randomly selected from the word "calculator." Find the probability that the letter is

   a. A "u".

   b. An "a".

   c. A "c" or an "l".

   d. A vowel.

7. There are four women and five men employed in a real estate office. If a person is selected at random to get lunch for the group, find the probability that the person is a woman.

8. A marble is selected at random from a bag containing two red marbles, one blue marble, three green marbles, and one white marble. Find the probability that the ball is

   a. A green marble.

   b. A red or a blue marble.

   c. An orange marble.

9. On a roulette wheel there are 38 sectors. Eighteen are red, 18 are black, and two are green. When the wheel is spun, find the probability that the ball will land on

   a. Black.

   b. Green.

10. A person has a penny, a nickel, a dime, a quarter, and a half-dollar in his pocket. If a coin is selected at random, find the probability that the coin is

    a. A nickel.

    b. A coin whose amount is greater than 10 cents.

    c. A coin whose denomination ends in a 0.

ANSWERS

1. The sample space is 1, 2, 3, 4, 5, 6.

   a. $P(4) = \frac{1}{6}$, since there is only one 4 in the sample space.

   b. $P(\text{number less than 3}) = \frac{2}{6} = \frac{1}{3}$, since there are 2 numbers in the sample space less than 3.

   c. P(number less than 1) $= \frac{0}{6} = 0$, since there are no numbers in the sample space less than 1.

   d. P(number is divisible by 3) $= \frac{2}{6} = \frac{1}{3}$, since 3 and 6 are divisible by 3.

2. The sample space is $1, $2, $5, $10, $20.

   a. P($2) $= \frac{1}{5}$.

   b. P(bill greater than $5) $= \frac{2}{5}$, since $10 and $20 are greater than $5.

   c. P($100) $= \frac{0}{5} = 0$, since there is no $100 bill in the sample space.

   d. P(bill is even) $= \frac{3}{5}$, since $2, $10, and $20 are even denomination bills.

   e. P(number is divisible by 10) $= \frac{2}{5}$, since $10 and $20 are divisible by 10.

3. The sample space is HH, HT, TH, TT.

   a. P(HH) $= \frac{1}{4}$, since there is only 1 way to get 2 heads.

   b. P(at least one tail) $= \frac{3}{4}$, since there are 3 ways (HT, TH, TT) to get at least 1 tail.

   c. P(TT) $= \frac{1}{4}$, since there is only 1 way to get 2 tails.

4. The sample space is A♥, 2♦, 3♣, 4♥, 5♠, 6♣.

   a. P(ace of hearts) $= \frac{1}{6}$.

   b. P(black card) $= \frac{3}{6} = \frac{1}{2}$, since there are 3 black cards.

   c. P(spade) $= \frac{1}{6}$, since there is 1 spade.

5. The sample space is 1, 2, 3, 4, 5.

   a. P(number divisible by 2) $= \frac{2}{5} = 5$, since 2 and 4 are divisible by 2.

   b. P(number greater than 6) $= \frac{0}{5} = 0$, since no numbers are greater than 6.

   c. P(odd number) $= \frac{3}{5}$, since 1, 3, and 5 are odd numbers.

6. The sample space consists of the letters in "calculator."

   a. P(u) $= \frac{1}{10}$.

   b. P(a) $= \frac{2}{10} = \frac{1}{5}$.

   c. P(c or l) $= \frac{4}{10} = \frac{2}{5}$.

   d. P(vowel) $= \frac{4}{10} = \frac{2}{5}$, since a, u, and o are vowels.

7. The sample space consists of four women and five men. P(woman) $= \frac{4}{9}$.

8. The sample space is red, red, blue, green, green, green, and white.

   a. P(green) $= \frac{3}{7}$, since there are 3 green marbles.

   b. P(red or blue) $= \frac{3}{7}$, since there are 2 red marbles and 1 blue marble.

   c. P(orange) $= \frac{0}{7} = 0$, since there are no orange marbles.

9. There are 38 outcomes.

  a. $P(\text{black}) = \frac{18}{38} = \frac{9}{19}$.

  b. $P(\text{green}) = \frac{2}{38} = \frac{1}{19}$.

10. The sample space is 1¢, 5¢, 10¢, 25¢, 50¢.

  a. $P(5¢) = \frac{1}{5}$.

  b. $P(\text{greater than } 10¢) = \frac{2}{5}$.

  c. $P(\text{denomination ends in } 0) = \frac{2}{5}$.

Empirical Probability

Probabilities can be computed for situations that do not use sample spaces. In these cases, **frequency distributions** are used and the probability is called **empirical probability**. For example, suppose a class of students consists of six freshmen, two sophomores, 10 juniors, and 12 seniors. The information can be summarized in a frequency distribution as follows:

Rank	Frequency
Freshmen	6
Sophomores	2
Juniors	10
Seniors	12
TOTAL	30

From a frequency distribution, probabilities can be computed using the formula

$$P(E) = \frac{\text{frequency of } E}{\text{sum of the frequencies}}$$

Empirical probability is sometimes called **relative frequency** probability.

**EXAMPLE**

Using the frequency distribution shown previously, find the probability of selecting a senior student at random.

**SOLUTION**

Since there are 12 seniors and a total of 30 students, $P(\text{senior}) = \frac{12}{30} = \frac{2}{5}$.

Another aspect of empirical probability is that if a large number of subjects (called a sample) is selected from a particular group (called a population), and the probability of a specific attribute is computed, then when another subject is selected, we can say the probability that this subject has the same attribute is the same as the original probability computed for the group. For example, a Gallup Poll of 1004 adults surveyed found that 17% of the subjects stated that they considered Abraham Lincoln to be the greatest President of the United States. Now if a subject is selected, the probability that he or she will say President Lincoln is the greatest president is 17%.

Several things should be explained here. First of all, the 1004 people constituted a sample selected from a larger group called the population. Second, the exact probability for the population can never be known unless every single member of the population is surveyed. This does not happen in these kinds of surveys because the population is usually very large. Hence, the 17% is only an estimate of the probability. However, if the sample is **representative** of the population, the estimate will usually be fairly close to the exact probability. Statisticians have a way of computing the accuracy (called the margin of error) for these situations. For the present, we shall just concentrate on the probability.

Also, by a representative sample, we mean the subjects of the sample have similar characteristics as those in the population. There are statistical methods to help the statisticians obtain a representative sample. These methods are called sampling methods and can be found in many statistics books.

EXAMPLE

The same study found 7% considered George Washington to be the greatest president. If a person is selected at random, find the probability that he or she considers George Washington to be the greatest president.

SOLUTION

The probability is 7%.

EXAMPLE

In a sample of 832 people over 30 years of age, 180 had a bachelor's degree. If a person over 30 years of age is selected, find the probability that the person has a bachelor's degree.

 **SOLUTION**

In this case,

$$P(\text{bachelor's degree}) = \tfrac{180}{832} = 0.216 \text{ (rounded) or about 21.6\%.}$$

 **EXAMPLE**

In the sample study of 832 people, it was found that 680 people have a high school diploma. If a person is selected at random, find the probability that the person does not have a high school diploma.

 **SOLUTION**

The probability that a person has a high school diploma is

$$P(\text{high school diploma}) = \tfrac{680}{832} = 0.817 \text{ (rounded)} = 81.7\%.$$

Hence, the probability that a person does not have a high school diploma is

$$P(\text{no high school diploma}) = 1 - P(\text{high school diploma})$$
$$= 1 - 0.817 = 0.183 = 18.3\%$$

Alternate Solution:

If 680 people have a high school diploma, then 832 − 680 = 152 do not have a high school diploma. Hence,

$$P(\text{no high school diploma}) = \tfrac{152}{832} = 0.183 \text{ (rounded) or 18.3\%.}$$

Consider another aspect of probability. Suppose a baseball player has a batting average of 0.250. What is the probability that he will get a hit the next time he gets to bat? Although we cannot be sure of the exact probability, we can use 0.250 as an estimate. Since $0.250 = \tfrac{1}{4}$, we can say that there is a one in four chance that he will get a hit the next time he bats.

 Still Struggling

When the probability of an event is expressed as a fraction with a large denominator, it might be more meaningful to represent it as a decimal or percent.

PRACTICE

1. In a sample of 60 people, 18 had type O blood, 32 had type A blood, 7 had type B blood, and 3 had type AB blood. If a person is selected at random, find the probability that the person

   a. Has type B blood

   b. Has type B or type AB blood

   c. Does not have type A blood

   d. Has neither type AB nor type O blood

2. In a recent survey of 425 children ages 6 to 8 years, it was found that 132 did not like to go to school. If a child is selected at random, find the probability that he or she did not like to go to school.

3. A recent survey found that the ages of nurses in a hospital network is distributed as follows:

Age	Number
20–29	10
30–39	36
40–49	22
50–59	18
60 or older	14
	100

   If a person is selected at random, find the probability that the person is

   a. 50 or older

   b. Under 30 years old

   c. Between 40 and 59 years old

   d. Under 60 but over 29 years old

4. A recent survey found that 64% of those questioned get some of the news from the Internet. If a person is selected at random, find the probability that the person does not get any news from the Internet.

5. In a classroom of 32 students, 12 were liberal arts majors and 18 were history majors. If a student is selected at random, find the probability that the student is neither a liberal arts major nor a history major.

**ANSWERS**

1. **The total number of outcomes in this sample space is 60.**

    a. $P(B) = \frac{7}{60}$

    b. $P(B \text{ or } AB) = \frac{7+3}{60} = \frac{10}{60} = \frac{1}{6}$

    c. $P(\text{not } A) = 1 - \frac{32}{60} = \frac{28}{60} = \frac{7}{15}$

    d. $P(\text{neither } AB \text{ nor } O) = P(A \text{ or } B) = \frac{32+7}{60} = \frac{39}{60} = \frac{13}{20}$

2. $P(\text{not like school}) = \frac{132}{425}$

3. a. $P(50 \text{ or older}) = \frac{18+14}{100} = \frac{32}{100} = \frac{8}{25}$

    b. $P(\text{under } 30) = \frac{10}{100} = \frac{1}{10}$

    c. $P(\text{between 40 and 59}) = \frac{22+18}{100} = \frac{40}{100} = \frac{2}{5}$

    d. $P(\text{over 29 but under 60}) = \frac{36+22+18}{100} = \frac{76}{100} = \frac{19}{25}$

4. $P(\text{no news from Internet}) = 1 - 0.64 = 0.36$

5. $P(\text{neither liberal arts nor history}) = 1 - \frac{12+18}{32} = 1 - \frac{30}{32} = \frac{2}{32} = \frac{1}{16}$

Law of Large Numbers

We know from classical probability that if a coin is tossed one time, we cannot predict the outcome, but the probability of getting a head is $\frac{1}{2}$ and the probability of getting a tail is $\frac{1}{2}$ if everything is fair. But what happens if we toss the coin 100 times? Will we get 50 heads? Common sense tells us that most of the time, we will not get exactly 50 heads, but we should get close to 50 heads. What will happen if we toss a coin 1000 times? Will we get exactly 500 heads? Probably not. However, as the number of heads increases, the ratio of the number of heads to the total number of tosses will get closer to $\frac{1}{2}$. This phenomenon is known as the **law of large numbers**. This law holds for any type of gambling game such as rolling dice, playing roulette, etc.

Subjective Probability

A third type of probability is called **subjective probability**. Subjective probability is based upon an educated guess, estimate, opinion, or inexact information. For example, a sports writer may say that there is a 60% probability that the Pittsburgh Steelers will be in the Super Bowl next year. Here the sports writer is basing his opinion on subjective information such as the relative strength of the Steelers, their opponents, their coach, etc. Two individuals' estimates of subjective probability can be different since each person would have different information about the situation. For example, a union president might guess the probability of the employees going on strike at 80%, whereas the owner of the company might say it would be 60%. Each person could be getting information from different sources. Subjective probabilities are used in everyday life; however, they are beyond the scope of this book.

PROBABILITY SIDELIGHT: Brief History of Probability

The concepts of probability are as old as humans. Paintings in tombs excavated in Egypt showed that people played games based on chance as early as 1800 BCE. One game was called "Hounds and Jackals" and is similar to the present-day game of "Snakes and Ladders."

Ancient Greeks and Romans made crude dice from various items such as animal bones, stones, and ivory. When some of these items were tested recently, they were found to be quite accurate. These crude dice were also used in fortune telling and divination.

Emperor Claudius (10 BCE–54 CE) is said to have written a book entitled *How To Win at Dice*. He liked to play so much that he had a special dice board in his carriage.

No formal study of probability was done until the 16th century when Girolamo Cardano (1501–1576) wrote a book on probability entitled *The Book on Chance and Games*. Cardano was a philosopher, astrologer, physician, mathematician, and gambler. In his book, he also included techniques on how to cheat and how to catch others who are cheating. He is believed to be the first mathematician to formulate a definition of classical probability.

During the mid-1600s, a professional gambler named Chevalier de Mere made a considerable amount of money on a gambling game. He would bet unsuspecting patrons that in four rolls of a die, he could obtain at least one 6. He was so successful at winning that word got around, and people refused to play. He decided to invent a new game in order to keep winning. He would bet patrons that if he rolled two dice 24 times, he would get at least one double 6. However, to his dismay, he began to lose more often than he would win.

Unable to figure out why he was losing, he asked a renowned mathematician, Blaise Pascal (1623–1662) to study the game. Pascal was a child prodigy when it came to mathematics. At the age of 14, he participated in weekly meetings of the mathematicians of the French Academy. At the age of 16, he invented a mechanical adding machine.

Because of the dice problem, Pascal became interested in studying probability and began a correspondence with a French government official and fellow mathematician, Pierre de Fermat (1601–1665). Together the two were able to solve de Mere's dilemma and formulate the beginnings of probability theory.

In 1657, a Dutch mathematician named Christian Huygens wrote a treatise on the Pascal–Fermat correspondence and introduced the idea of mathematical expectation. (See Chapter 5.)

Abraham de Moivre (1667–1754) wrote a book on probability entitled *Doctrine of Chances* in 1718. He published a second edition in 1738.

Pierre Simon Laplace (1749–1827) wrote a book and a series of supplements on probability from 1812 to 1825. His purpose was to acquaint readers with the theory of probability and its applications using everyday language. He also stated that the probability that the sun will rise tomorrow is $\frac{1,826,214}{1,826,215}$.

Simeon-Denis Poisson (1781–1840) developed the concept of the Poisson distribution. (See Chapter 8.)

Also during the 1800s, a mathematician named Carl Friedrich Gauss (1777–1855) developed the concepts of the normal distribution. Earlier work on the normal distribution was also done by DeMoivre and Laplace, unknown to Gauss. (See Chapter 9.)

In 1895, the Fey Manufacturing Company of San Francisco invented the first automatic slot machine. These machines consisted of three wheels that were spun when a handle on the side of the machine was pulled. Each wheel contained 20 symbols; however, the number of the same symbols was not the same on each wheel. For example, the first wheel may have 6 oranges, while the second wheel has 3 oranges, and the third wheel has 1 orange. When a person gets two oranges, the person may think that he has almost won by getting 2 out of 3 equitable symbols when the real probability of winning is much smaller.

In the late 1940s, two mathematicians, John von Neumann and Stanislaw Ulam, used a computer to simulate probability experiments. This method is called the Monte Carlo method. (See Chapter 10.)

Today probability theory is used in insurance, gambling, war gaming, the stock market, weather predicting, and many other areas.

Summary

Probability is the mathematics of chance. There are three types of probability: classical probability, empirical probability, and subjective probability. Classical probability uses sample spaces. A sample space is the set of outcomes of a probability experiment. The range of probability is from 0 to 1. If an event cannot occur, its probability is 0. If an event is certain to occur, its probability is 1. Classical probability is defined as the number of ways (outcomes) the event can occur divided by the total number of outcomes in the sample space.

Empirical probability uses frequency distributions, and it is defined as the frequency of an event divided by the total number of frequencies.

Subjective probability is made by a person's knowledge of the situation and is basically an educated guess as to the chances of an event occurring.

QUIZ

1. **Rolling a die or tossing a coin is called a**
   A. Repeated experiment
   B. Sample experiment
   C. Probability experiment
   D. Infinite experiment

2. **Which is not a type of probability?**
   A. Subjective
   B. Classical
   C. Variable
   D. Empirical

3. **The range of the values a probability can assume is**
   A. From 0 to 1
   B. From 1 to 100
   C. From 0 to $\frac{1}{2}$
   D. From -1 to $+1$

4. **When an event cannot occur, its probability is**
   A. $\frac{1}{2}$
   B. 0.01
   C. 1
   D. 0

5. **The set of all possible outcomes of a probability experiment is called the**
   A. Outcome space
   B. Sample space
   C. Experimental space
   D. Event space

6. **How many outcomes are there altogether when three coins are tossed?**
   A. 1
   B. 2
   C. 4
   D. 8

7. **The type of probability that uses sample spaces is called**
   A. Subjective probability
   B. Empirical probability
   C. Classical probability
   D. Relative probability

8. When an event is certain to occur, its probably is
   A. −1
   B. $\frac{1}{2}$
   C. 0
   D. 1

9. When two coins are tossed, the sample space is
   A. H, T, HT
   B. HH, HT, TT
   C. H, T
   D. HH, HT, TH, TT

10. When a die is rolled, the probability of getting a number greater than 2 is
    A. $\frac{1}{6}$
    B. $\frac{1}{3}$
    C. $\frac{1}{2}$
    D. $\frac{2}{3}$

11. When two coins are tossed, the probability of getting two heads is
    A. $\frac{1}{3}$
    B. $\frac{1}{2}$
    C. $\frac{1}{8}$
    D. $\frac{1}{4}$

12. If a letter is selected at random from the word "Mississippi," find the probability that it is an "i."
    A. $\frac{1}{2}$
    B. $\frac{4}{11}$
    C. $\frac{1}{8}$
    D. $\frac{3}{11}$

13. When a die is rolled, the probability of getting a 7 is
    A. 0
    B. $\frac{1}{6}$
    C. $\frac{1}{2}$
    D. 1

14. In a survey of 240 people, 95 were over the age of 62. If a person is selected at random, what is the probability that the person is over 62?

   A. $\frac{16}{45}$

   B. $\frac{19}{48}$

   C. $\frac{37}{90}$

   D. $\frac{53}{90}$

15. In a classroom of 27 students, there were 21 freshmen. If a student is selected at random, what is the probability that the student is not a freshman?

   A. $\frac{7}{9}$

   B. $\frac{4}{9}$

   C. $\frac{2}{9}$

   D. $\frac{5}{9}$

chapter 2

Sample Spaces

In order to compute classical probabilities, you need to find the sample space for a probability experiment. In the previous chapter, sample spaces were found by using common sense. In this chapter, two specific devices can be used to find sample spaces for probability experiments: tree diagrams and tables.

CHAPTER OBJECTIVES

In this chapter, you will learn

- How to find a sample space using a tree diagram
- How to find a sample space using a table

Tree Diagrams

A **tree diagram** consists of branches corresponding to the outcomes of two or more probability experiments that are done in sequence.

In order to construct a tree diagram, use branches corresponding to the outcomes of the first experiment. These branches will emanate from a single point. Then from each branch of the first experiment draw branches that represent the outcomes of the second experiment. You can continue the process for further experiments of the sequence if necessary.

 EXAMPLE

A coin is tossed and a die is rolled. Draw a tree diagram and find the sample space.

 **SOLUTION**

a. Since there are two outcomes (heads and tails for the coin), draw two branches from a single point and label one H for head and the other one T for tail.

b. From each one of these outcomes, draw and label six branches representing the outcomes 1, 2, 3, 4, 5, and 6 for the die.

c. Trace through each branch to find the outcomes of the experiment. See Figure 2-1.

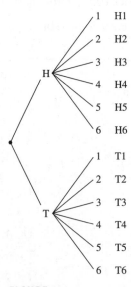

FIGURE 2-1

Hence, there are 12 outcomes. They are H1, H2, H3, H4, H5, H6, T1, T2, T3, T4, T5, T6.

Once the sample space has been found, probabilities for events can be computed.

 **EXAMPLE**

A coin is tossed and a die is rolled. Find the probability of getting
  a. A tail on the coin and a 5 on the die
  b. A tail on the coin
  c. A 3 on the die

 **SOLUTION**

  a. Since there are 12 outcomes in the sample space and only one way to get a tail on the coin and a 5 on the die,

$$P(T5) = \frac{1}{12}.$$

  b. Since there are six ways to get a tail on the coin, namely T1, T2, T3, T4, T5, and T6,

$$P(\text{tail on the coin}) = \frac{6}{12} = \frac{1}{2}.$$

  c. Since there are two ways to get a 3 on the die, namely H3 and T3,

$$P(\text{3 on the die}) = \frac{2}{12} = \frac{1}{6}.$$

 **EXAMPLE**

Three coins are tossed. Draw a tree diagram and find the sample space.

 SOLUTION

Each coin can land either heads up (H) or tails up (T); therefore, the tree diagram will consist of three parts and each part will have two branches. See Figure 2-2.

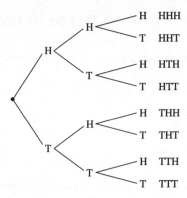

FIGURE 2-2

Hence, the sample space is HHH, HHT, HTH, HTT, THH, THT, TTH, TTT.

Once the sample space is found, probabilities can be computed.

EXAMPLE

Three coins are tossed. Find the probability of getting
a. Two tails and a head in any order

b. Three tails

c. No tails

d. At least two heads

e. At most two heads

**SOLUTION**

a. **There are eight outcomes in the sample space, and there are three ways to get two tails and a head in any order. They are TTH, THT, and HTT; hence,**

$$P(\text{two tails and a head in any order}) = \frac{3}{8}.$$

b. **Three tails can occur in only one way; hence,**

$$P(\text{TTT}) = \frac{1}{8}.$$

c. **The event of getting no tails can occur in only one way—namely HHH; hence,**

$$P(\text{HHH}) = \frac{1}{8}.$$

d. The event of at least two heads means two heads and one tail in any order or three heads. There are four outcomes in this event—namely HHT, HTH, THH, and HHH; hence,

$$P(\text{at least 2 heads}) = \frac{4}{8} = \frac{1}{2}.$$

e. The event of getting at most two heads means zero heads, one head, or two heads. There are seven outcomes in this event—namely TTT, HTT, THT, TTH, HHT, HTH, and THH; hence,

$$P(\text{at most 2 heads}) = \frac{7}{8}.$$

When selecting more than one object from a group of objects, it is important to know whether or not the object selected is replaced before drawing the second object. Consider the next two examples.

EXAMPLE

A box contains a red ball (R), a blue ball (B), and a yellow ball (Y). Two balls are selected at random in succession. Draw a tree diagram and find the sample space if the first ball is **replaced** before the second ball is selected.

SOLUTION

There are three ways to select the first ball. They are a red ball, a blue ball, or a yellow ball. Since the first ball is replaced before the second one is selected, there are three ways to select the second ball. They are a red ball, a blue ball, or a yellow ball. The tree diagram is shown in Figure 2-3.

The sample space consists of nine outcomes. They are RR, RB, RY, BR, BB, BY, YR, YB, YY. Each outcome has a probability of $\frac{1}{9}$.

Now what happens if the first ball is not replaced before the second ball is selected?

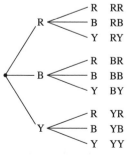

FIGURE 2-3

EXAMPLE

A box contains a red ball (R), a blue ball (B), and a yellow ball (Y). Two balls are selected at random in succession. Draw a tree diagram and find the sample space if the first ball is **not replaced** before the second ball is selected.

 SOLUTION

There are three outcomes for the first ball. They are a red ball, a blue ball, or a yellow ball. Since the first ball is not replaced before the second ball is drawn, there are only two outcomes for the second ball, and these outcomes depend on the color of the first ball selected. If the first ball selected is blue, then the second ball can be either red or yellow, etc. The tree diagram is shown in Figure 2-4.

The sample space consists of six outcomes, which are RB, RY, BR, BY, YR, YB. Each outcome has a probability of $\frac{1}{6}$.

? Still Struggling

Tree diagrams can only be used to find sample spaces for probability experiments with a limited number of outcomes. There are other methods, such as permutations and combinations (shown in a later chapter), that will give the number of outcomes in a probability experiment.

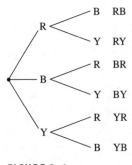

FIGURE 2-4

PRACTICE

1. If the possible blood types are A, B, AB, and O, and each type can be Rh⁺ or Rh⁻, draw a tree diagram and find all possible blood types.

2. Students are classified as male (M) or female (F), freshman (Fr), sophomore (So), junior (Jr), or senior (Sr), and full-time (Ft) or part-time (Pt). Draw a tree diagram and find all possible classifications.

3. A box contains a $1 bill, a $5 bill, and a $10 bill. Two bills are selected in succession **with** replacement. Draw a tree diagram and find the sample space. Find the probability that the total amount of money selected is

   a. $11

   b. Greater than $6

   c. Less than $10

4. Draw a tree diagram and find the sample space for the genders of the children in a family consisting of three children. Assume the genders are equiprobable. Find the probability of

   a. Three boys

   b. Two girls and a boy in any order

   c. At least two girls

5. A box contains a white marble (W), a blue marble (B), and a green marble (G). Two marbles are selected **without** replacement. Draw a tree diagram and find the sample space. Find the probability that one marble is blue.

ANSWERS

1.

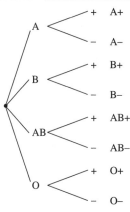

FIGURE 2-5

2.

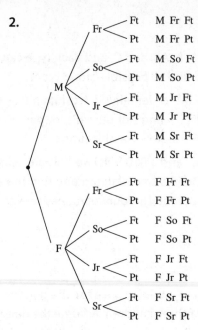

FIGURE 2-6

3.

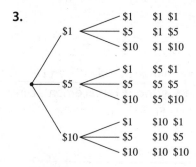

FIGURE 2-7

There are nine outcomes in the sample space

a. $P(\$11) = \frac{2}{9}$, since $1, $10 and $10, $1 equal $11.

b. $P(\text{greater than } \$6) = \frac{6}{9} = \frac{2}{3}$, since there are 6 ways to get a sum greater than $6.

c. $P(\text{less than } \$10) = \frac{3}{9} = \frac{1}{3}$, since there are 3 ways to get a sum less than $10.

4.

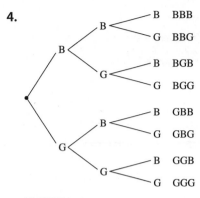

FIGURE 2-8

There are eight outcomes in the sample space

a. $P(3 \text{ boys}) = \frac{1}{8}$, since 3 boys is BBB.

b. $P(2 \text{ girls and one boy in any order}) = \frac{3}{8}$, since there are 3 ways to get 2 girls and one boy in any order. They are GGB, GBG, and BGG.

c. $P(\text{at least 2 girls}) = \frac{4}{8} = \frac{1}{2}$, since at least 2 girls means 2 or 3 girls. The outcomes are GGB, GBG, BGG, and GGG.

5.

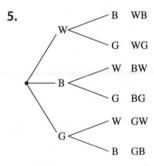

FIGURE 2-9

The probability that one marble is blue is $\frac{4}{6} = \frac{2}{3}$, since the outcomes are WB, BW, BG, and GB.

Tables

Another way to find a sample space is to use a table.

 EXAMPLE

Find the sample space for selecting a card from a standard deck of 52 cards.

 SOLUTION

There are four suits—namely hearts and diamonds, which are red, and spades and clubs, which are black. Each suit consists of 13 cards—namely ace through king. Hence, the sample space can be shown using a table. See Figure 2-10.

Face cards are kings, queens, and jacks.

Once the sample space is found, probabilities for events can be computed.

 EXAMPLE

A single card is drawn at random from a standard deck of cards. Find the probability that it is

   a. The 3 of clubs

   b. A jack

   c. A 7 or a diamond

   d. A 7 and a diamond

 SOLUTION

   a. The sample space consists of 52 outcomes and only one outcome is the 3 of clubs; hence,

$$P(3\clubsuit) = \frac{1}{52}.$$

A	2	3	4	5	6	7	8	9	10	J	Q	K
♥	♥	♥	♥	♥	♥	♥	♥	♥	♥	♥	♥	♥

A	2	3	4	5	6	7	8	9	10	J	Q	K
♦	♦	♦	♦	♦	♦	♦	♦	♦	♦	♦	♦	♦

A	2	3	4	5	6	7	8	9	10	J	Q	K
♠	♠	♠	♠	♠	♠	♠	♠	♠	♠	♠	♠	♠

A	2	3	4	5	6	7	8	9	10	J	Q	K
♣	♣	♣	♣	♣	♣	♣	♣	♣	♣	♣	♣	♣

FIGURE 2-10

b. Since there are four jacks (one of each suit)

$$P(J) = \frac{4}{52} = \frac{1}{13}.$$

c. In this case, there are 13 diamonds and four 7s; however, the 7♦ has been counted twice so the number of ways to get a 7 or a diamond is $13 + 4 - 1 = 16$. Hence,

$$P(7 \text{ or } ♦) = \frac{16}{52} = \frac{4}{13}.$$

d. Since there is only one way to get a 7 and a diamond, that is, the 7 of diamonds,

$$P(7♦) = \frac{1}{52}.$$

A table can be used for the sample space when two dice are rolled. Since the first die can land in six ways and the second die can land in six ways, there are 6×6 or 36 outcomes in the sample space. It does not matter whether the two dice are the same color or whether they are different colors. The sample space is shown in Figure 2-11.

Notice the sample space consists of ordered pairs of numbers. The outcome (4, 2) means that a 4 was obtained on die one and a 2 was obtained on die two. The sum of the spots on the faces in this case is $4 + 2 = 6$. Probability problems involving rolling two dice can be solved using the sample space shown in Figure 2-11.

Die 2	Die 1					
	1	2	3	4	5	6
1	(1, 1)	(2, 1)	(3, 1)	(4, 1)	(5, 1)	(6, 1)
2	(1, 2)	(2, 2)	(3, 2)	(4, 2)	(5, 2)	(6, 2)
3	(1, 3)	(2, 3)	(3, 3)	(4, 3)	(5, 3)	(6, 3)
4	(1, 4)	(2, 4)	(3, 4)	(4, 4)	(5, 4)	(6, 4)
5	(1, 5)	(2, 5)	(3, 5)	(4, 5)	(5, 5)	(6, 5)
6	(1, 6)	(2, 6)	(3, 6)	(4, 6)	(5, 6)	(6, 6)

FIGURE 2-11

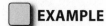 **EXAMPLE**

When two dice are rolled, find the probability of getting a sum of 10.

SOLUTION

There are three ways of rolling a sum of ten. They are (6, 4), (5, 5), and (4, 6). The sample space consists of 36 outcomes. Hence,

$$P(\text{sum of 10}) = \frac{3}{36} = \frac{1}{12}.$$

EXAMPLE

When two dice are rolled, find the probability of getting doubles.

SOLUTION

There are six ways to get doubles. They are (1, 1), (2, 2), (3, 3), (4, 4), (5, 5), and (6, 6); hence,

$$P(\text{doubles}) = \frac{6}{36} = \frac{1}{6}.$$

EXAMPLE

When two dice are rolled, find the probability of getting a sum less than 4.

SOLUTION

A sum less than 4 means a sum of 3 or 2. There are two ways of getting a sum of 3. They are (2, 1) and (1, 2). There is one way of getting a sum of 2. It is (1, 1). The total number of ways of getting a sum less than 4 is 2 + 1 = 3. Hence,

$$P(\text{sum less than 4}) = \frac{3}{36} = \frac{1}{12}.$$

EXAMPLE

When two dice are rolled, find the probability that one or both of the numbers is a 4.

SOLUTION

There are 11 outcomes that contain a 4. They are (1, 4), (2, 4), (3, 4), (4, 4), (5, 4), (6, 4), (4, 1), (4, 2), (4, 3), (4, 5), and (4, 6). Hence,

$$P(\text{one of the numbers is a 4}) = \frac{11}{36}.$$

Still Struggling

Remember, when two dice are rolled, there are 36 outcomes in the sample space. They consist of ordered pairs of numbers. The sums of the faces of the dice are 2 through 12 and are not the outcomes of the sample space.

PRACTICE

1. When a card is selected at random from a deck, find the probability of getting

   a. A spade

   b. The king of clubs

   c. A 10

2. When a card is selected at random from a deck, find the probability of getting

   a. A red card

   b. A black king

   c. A club or a diamond

3. When a card is selected at random from a deck, find the probability of getting

   a. A club or a face card

   b. A diamond or a 6

   c. A black face card

4. When two dice are rolled, find the probability of getting

   a. A sum of 11

   b. A sum greater than 5

   c. A sum less than or equal to 4

5. When two dice are rolled, find the probability of getting

   a. A 6 on one or both dice

   b. A sum less than 13

   c. A sum greater than 12

ANSWERS

1. There are 52 outcomes in the sample space.

    a. There are 13 spades, so

    $$P(\spadesuit) = \frac{13}{52} = \frac{1}{4}.$$

    b. There is only one king of clubs, so

    $$P(K\clubsuit) = \frac{1}{52}.$$

    c. There are four 10s, so

    $$P(10) = \frac{4}{52} = \frac{1}{13}.$$

2. There are 52 outcomes in the sample space.

    a. There are 26 red cards—that is, 13 diamonds and 13 hearts, so

    $$P(\text{red card}) = \frac{26}{52} = \frac{1}{2}.$$

    b. There are two black kings: they are the king of spades and the king of clubs, so

    $$P(\text{black king}) = \frac{2}{52} = \frac{1}{26}.$$

    c. There are 13 clubs and 13 diamonds, so

    $$P(\clubsuit \text{ or } \blacklozenge) = \frac{26}{52} = \frac{1}{2}.$$

3. There are 52 outcomes in the sample space.

    a. There are 13 clubs and 12 face cards, but the jack, queen, and king of clubs have been counted twice, so

    $$P(\text{club or face card}) = \frac{13+12-3}{52} = \frac{22}{52} = \frac{11}{26}.$$

    b. There are 13 diamonds and 4 sixes, but the 6 of diamonds has been counted twice, so

    $$P(\clubsuit \text{ or an } 8) = \frac{13+4-1}{52} = \frac{16}{52} = \frac{4}{13}.$$

    c. There are 12 face cards, and $\frac{1}{2}$ of them are black

    $$P(\text{black face card}) = \frac{6}{52} = \frac{3}{26}.$$

4. There are 36 outcomes in the sample space.

   a. There are two ways to get a sum of 11. They are (5, 6) and (6, 5); hence,

   $$P(\text{sum of 11}) = \frac{2}{36} = \frac{1}{18}.$$

   b. A sum greater than 5 means a sum of 6, 7, 8, 9, 10, 11, or 12.

   $$P(\text{sum greater than 5}) = \frac{26}{36} = \frac{13}{18}.$$

   c. A sum less than or equal to 4 means a sum of 4, 3, or 2. There are six ways to get a sum less than or equal to 4; hence,

   $$P(\text{sum less than or equal to 4}) = \frac{6}{36} = \frac{1}{6}.$$

5. There are 36 outcomes in the sample space.

   a. There are 11 ways to get a 6 on one or both dice. They are (1, 6), (2, 6), (3, 6), (4, 6), (5, 6), (6, 6), (6, 5), (6, 4), (6, 3), (6, 2), and (6, 1); hence,

   $$P(\text{6 on one or both dice}) = \frac{11}{36}.$$

   b. Since all sums are less than 13 when two dice are rolled, there are 36 ways to get a sum less than 13; hence,

   $$P(\text{sum less than 13}) = \frac{36}{36} = 1.$$

   The event is certain.

   c. There are zero ways to get a sum greater than 12; hence,

   $$P(\text{sum greater than 12}) = \frac{0}{36} = 0.$$

   The event is impossible.

PROBABILITY SIDELIGHT: History of Dice and Cards

Dice are one of the earliest known gambling devices used by humans. They have been found in ancient Egyptian tombs and in the prehistoric caves of people in Europe and America. The first dice were made from animal bones—namely the astragalus or the heel bone of a hoofed animal. These bones are very smooth and easily carved. The astragalus had only four sides as opposed to modern cubical dice that have six sides. The astragalus were used for fortune telling, gambling, and board games.

By 3000 BCE, the Egyptians had devised many board games. Ancient tomb paintings show pharaohs playing board games, and a game similar to today's "Snakes and Ladders" was found in an Egyptian tomb dating from around 1800 BCE. Eventually crude cubic dice evolved from the astragalus. The dice were first made from bones, then clay, wood, and finally polished stones. Dots were used instead of numbers since writing numbers was very complicated at that time.

It was thought that the outcomes of rolled dice were controlled by the gods that the people worshipped. As one story goes, the Romans incorrectly reasoned that there were three ways to get a sum of 7 when two dice are rolled. They are 6 and 1, 5 and 2, and 4 and 3. They also reasoned incorrectly that there were three ways to get a sum of 6: 5 and 1, 4 and 2, and 3 and 3. They knew from gambling with dice that a sum of 7 appeared more often than a sum of 6. They believed that the reason was that the gods favored the number 7 over the number 6 since 7 at the time was considered a "lucky number." Furthermore, they even "loaded" dice so that the faces showing 1 and 6 occurred more often than other faces would if the dice were fair.

It is interesting to note that on today's dice, the numbers on the opposite faces sum to 7. That is, 4 is opposite 3, 2 is opposite 5, and 6 is opposite 1. This was not always true. Early dice showed 1 opposite 2, 3 opposite 4, and 5 opposite 6. The changeover to modern configuration is believed to have occurred in Egypt.

Many of the crude dice have been tested and found to be quite accurate. Actually mathematicians did not begin to study the outcomes of dice until around the 16th century. The great astronomer Galilei Galileo is usually given credit for figuring out that when three dice are rolled, there are 216 total outcomes, and that a sum of 10 and 11 is more probable than a sum of 9 and 12. This fact was known intuitively by gamblers long before this time.

Today dice are used in many types of gambling games and many types of board games. Where would we be today without the game of Monopoly?

It is thought that playing cards evolved from long wooden sticks that had various markings and were used by early fortunetellers and gamblers in the Far East. When the Chinese invented paper over 2000 years ago, people marked long thin strips of paper and used them instead of wooden sticks.

Paper "cards" first appeared in Europe around 1300 and were widely used in most of the European countries. Some decks contained 17 cards; others had 22 cards. The early cards were hand-painted and quite expensive to produce. Later, stencils were used to cut costs.

Various markings on the cards changed quite often. Besides the four commonly used suits used today, early decks of cards had five or six suits and used other symbols such as coins, flowers, and leaves.

The first cards to be manufactured in the United States were made by Jazaniah Ford in the late 1700s. His company lasted over 50 years. The first book on gambling published in the United States was an edition of *Hoyle's Games*, which was printed in 1796.

Summary

Two devices can be used to find sample spaces. They are tree diagrams and tables.

A tree diagram consists of branches corresponding to the outcomes of two or more probability experiments that are done in sequence.

Sample spaces can also be found by using tables. For example, the outcomes when selecting a card from an ordinary deck can be represented by a table. The columns are the denominators of the cards, ace through king, and the rows are the suits of the cards. When two dice are rolled, the 36 outcomes can be represented by using a table. The columns represent the numbers 1 through 6 obtained on one die and the rows are the numbers obtained on the other die. Once a sample space is found, probabilities can be computed for specific events.

QUIZ

1. When a coin is tossed and a die is rolled, the probability of getting a head or an even number on the die is
    A. $\frac{1}{3}$
    B. $\frac{2}{3}$
    C. $\frac{3}{4}$
    D. $\frac{5}{6}$

2. When a coin is tossed and then a die is rolled, the probability of getting a head on the coin and a 3 on the die is
    A. $\frac{1}{4}$
    B. $\frac{1}{12}$
    C. $\frac{3}{4}$
    D. $\frac{1}{12}$

3. When three coins are tossed, the probability of getting one or more heads is
    A. $\frac{3}{8}$
    B. $\frac{1}{8}$
    C. $\frac{1}{2}$
    D. $\frac{7}{8}$

4. A box contains a penny, a nickel, a dime, and a quarter. If two coins are selected without replacement, the probability of getting an amount greater than 15¢ is
    A. $\frac{1}{2}$
    B. $\frac{2}{3}$
    C. $\frac{3}{4}$
    D. $\frac{5}{6}$

5. When three coins are tossed, the probability of getting at least two tails is
    A. $\frac{3}{8}$
    B. $\frac{1}{8}$
    C. $\frac{1}{2}$
    D. $\frac{5}{8}$

6. A bag contains a red bead, a green bead, a blue bead, and a yellow bead. If a bead is selected and its color noted, and then it is replaced and another bead is selected, the probability that both beads will be the same color is

   A. $\frac{1}{8}$

   B. $\frac{3}{4}$

   C. $\frac{1}{16}$

   D. $\frac{1}{4}$

7. A card is selected at random from a deck of 52 cards. The probability that it is a red 6 is

   A. $\frac{1}{52}$

   B. $\frac{1}{4}$

   C. $\frac{1}{26}$

   D. $\frac{1}{13}$

8. A card is selected at random from an ordinary deck of 52 cards. The probability that the 4 of clubs is selected is

   A. $\frac{1}{52}$

   B. $\frac{1}{13}$

   C. $\frac{1}{4}$

   D. $\frac{1}{26}$

9. A card is drawn from an ordinary deck of 52 cards. The probability that it is a red face card is

   A. $\frac{3}{52}$

   B. $\frac{9}{52}$

   C. $\frac{9}{13}$

   D. $\frac{3}{26}$

10. A card is drawn from an ordinary deck of 52 cards. The probability that it is a diamond is

    A. $\frac{1}{4}$

    B. $\frac{1}{52}$

    C. $\frac{1}{13}$

    D. $\frac{1}{26}$

11. A card is drawn from an ordinary deck of 52 cards. The probability that it is a 6 or a club is

   A. $\frac{3}{4}$

   B. $\frac{4}{13}$

   C. $\frac{5}{8}$

   D. $\frac{17}{52}$

12. Two dice are rolled. The probability that the sum of the spots on the faces will be 4 is

   A. $\frac{3}{13}$

   B. $\frac{1}{9}$

   C. $\frac{1}{12}$

   D. $\frac{5}{36}$

13. Two dice are rolled. The probability that the sum of the spots on the faces will be odd is

   A. $\frac{1}{2}$

   B. $\frac{3}{4}$

   C. $\frac{5}{6}$

   D. $\frac{3}{4}$

14. Two dice are rolled. The probability that the sum of the spots on the faces is greater than 6 is

   A. $\frac{7}{12}$

   B. $\frac{2}{3}$

   C. $\frac{5}{12}$

   D. $\frac{1}{2}$

15. Two dice are rolled. The probability that one or both numbers on the faces will be 2 is

   A. $\frac{1}{6}$

   B. $\frac{11}{36}$

   C. $\frac{1}{3}$

   D. $\frac{4}{13}$

chapter 3

The Addition Rules

In this chapter, the theory of probability is extended by using what are called the **addition rules**. Here one is interested in finding the probability of one event **or** another event occurring. In these situations, you must consider whether or not both events have common outcomes. For example, if you are asked to find the probability that you will get three oranges or three cherries on a slot machine, you know that these two events cannot occur at the same time if the machine has only three windows. In another situation you may be asked to find the probability of getting an odd number or a number less than 500 on a daily 3-digit lottery drawing. Here the events have common outcomes. For example, the number 451 is an odd number and a number less than 500. The two addition rules will enable you to solve these kinds of problems as well as many other probability problems.

CHAPTER OBJECTIVES

In this chapter, you will learn

- When two events are mutually exclusive
- How to find the probability of two events using the addition rules

Mutually Exclusive Events

Many problems in probability involve finding the probability of two or more events. For example, when a card is selected at random from a deck, what is the probability that the card is a king or a queen? In this case, there are two situations to consider. They are:

1. The card selected is a king.
2. The card selected is a queen.

Now consider another example. When a card is selected from a deck, find the probability that the card is a king or a diamond.

In this case, there are three situations to consider:

1. The card is a king.
2. The card is a diamond.
3. The card is a king and a diamond. That is, the card is the king of diamonds.

The difference is that in the first example, a card cannot be both a king and a queen at the same time, whereas in the second example, it is possible for the card selected to be a king and a diamond at the same time. In the first example, we say the two events are **mutually exclusive**. In the second example, we say the two events are **not** mutually exclusive. Two events then are **mutually exclusive** if they cannot occur at the same time; in other words, the events have no common outcomes.

EXAMPLE

Which events are mutually exclusive?
   a. Selecting a student who is a male or a senior
   b. Rolling two dice and getting a sum of 6 or 10
   c. Selecting a student at random who is a freshman or sophomore
   d. Selecting a card at random from a deck and getting a king or a heart
   e. Rolling a die and getting an even number or a number less than 3

**SOLUTION**
   a. No. A male student who is a senior is a common outcome.
   b. Yes.

c. Yes.

d. No. The king of hearts could be selected.

e. No. Two is a common outcome.

Addition Rule I

The probability of two or more events occurring can be determined by using the **addition rules**. The first rule is used when the events are mutually exclusive.

Addition Rule I: When two events are mutually exclusive,

$$P(A \text{ or } B) = P(A) + P(B).$$

EXAMPLE

A card is selected at random from a deck. Find the probability that the card is a queen or a 7.

SOLUTION

There are four queens and four 7s.

$$P(\text{queen or } 7) = P(\text{queen}) + P(7) = \frac{4}{52} + \frac{4}{52} = \frac{8}{52} = \frac{2}{13}.$$

EXAMPLE

When a die is rolled, find the probability of getting a 4 or a 5.

SOLUTION

As shown in Chapter 1, the problem can be done by looking at the sample space. It is 1, 2, 3, 4, 5, 6. Since there are two outcomes from six outcomes, $P(4 \text{ or } 5) = \frac{2}{6} = \frac{1}{3}$. Since the events are mutually exclusive, Addition Rule 1 also can be used:

$$P(4 \text{ or } 5) = P(4) + P(5) = \frac{1}{6} + \frac{1}{6} = \frac{2}{6} = \frac{1}{3}.$$

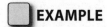

EXAMPLE

In a committee meeting, there were three freshmen, two sophomores, five juniors, and three seniors. If a student is selected at random to be the chairperson, find the probability that the chairperson is a sophomore or junior.

SOLUTION

There are two sophomores and five juniors and a total of 13 students.

$$P(\text{sophomore or junior}) = P(\text{sophomore}) + P(\text{junior}) = \frac{2}{13} + \frac{5}{13} = \frac{7}{13}.$$

Still Struggling

The word **or** is the key word, and it means one event occurs or the other event occurs.

PRACTICE

1. In a fish tank, there are 18 goldfish, six angelfish, and three guppies. If a person selects a fish at random, find the probability that the fish is

   a. A goldfish or an angelfish

   b. A guppy or an angelfish

2. A cable television system had eight horror movies, 12 drama movies, six mystery movies, and 14 comedy movies available to watch on a particular evening. If a person selects a movie at random, find the probability that it is

   a. A drama or a mystery movie

   b. A horror or a comedy movie

3. A bag of candy contains 20 root beer pieces, 18 caramel pieces, 15 peppermint pieces, and six lemon pieces. If a person selects a piece of candy at random, find the probability that it is

   a. A caramel or a peppermint piece

   b. A peppermint or a root beer piece

4. At a teachers' department meeting, there are six mathematics professors, eight computer science professors, and four statistics professors. If a professor is selected at random to conduct the meeting, find the probability that the professor is either

   a. A mathematics professor or a statistics professor

   b. A computer science professor or a statistics professor

5. On a particular day, a business receives 20 pieces of mail, eight are first class letters, nine are ads, and three are magazines. If a piece of mail is selected at random, find the probability that it is either

   a. An ad or a magazine

   b. A first class letter or an ad

ANSWERS

1. a. $P(\text{goldfish or angelfish}) = P(\text{goldfish}) + P(\text{angelfish}) = \frac{18}{27} + \frac{6}{27} = \frac{24}{27} = \frac{8}{9}$

   b. $P(\text{guppy or angelfish}) = P(\text{guppy}) + P(\text{angelfish}) = \frac{3}{27} + \frac{6}{27} = \frac{9}{27} = \frac{1}{3}$

2. a. $P(\text{drama or mystery}) = P(\text{drama}) + P(\text{mystery}) = \frac{12}{40} + \frac{6}{40} = \frac{18}{40} = \frac{9}{20}$

   b. $P(\text{horror or comedy}) = P(\text{horror}) + P(\text{comedy}) = \frac{8}{40} + \frac{14}{40} = \frac{22}{40} = \frac{11}{20}$

3. a. $P(\text{caramel or peppermint}) = P(\text{caramel}) + P(\text{peppermint}) = \frac{18}{59} + \frac{15}{59} = \frac{33}{59}$

   b. $P(\text{peppermint or root beer}) = P(\text{peppermint}) + P(\text{root beer}) = \frac{15}{59} + \frac{20}{59} = \frac{35}{59}$

4. a. $P(\text{mathematics professor or statistics professor}) = P(\text{mathematics professor}) + P(\text{statistics professor}) = \frac{6}{18} + \frac{4}{18} = \frac{10}{18} = \frac{5}{9}$

   b. $P(\text{computer science professor or statistics professor}) = P(\text{computer science professor}) + P(\text{statistics professor}) = \frac{8}{18} + \frac{4}{18} = \frac{12}{18} = \frac{2}{3}$

5. a. $P(\text{ad or magazine}) = P(\text{ad}) + P(\text{magazine}) = \frac{9}{20} + \frac{3}{20} = \frac{12}{20} = \frac{3}{5}$

   b. $P(\text{letter or ad}) = P(\text{letter}) + P(\text{ad}) = \frac{8}{20} + \frac{9}{20} = \frac{17}{20}$

Addition Rule II

When two events are not mutually exclusive, you need to add the probabilities of each of the two events and subtract the probability of the outcomes that are common to both events. In this case, Addition Rule II can be used.

Addition Rule II: If A and B are two events that are not mutually exclusive, then $P(A \text{ or } B) = P(A) + P(B) - P(A \text{ and } B)$, where A and B means the number of outcomes event A and event B have in common.

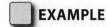 **EXAMPLE**

A card is selected at random from a deck of 52 cards. Find the probability that it is a 7 or a club.

 SOLUTION

Let A = the event of getting a 7; then $P(A) = \frac{4}{52}$ since there are four 7s. Let B = the event of getting a club; then $P(B) = \frac{13}{52}$ since there are 13 clubs. Since there is one card that is both a 7 and a club (i.e., the 7 of clubs), $P(A \text{ and } B) = \frac{1}{52}$. Hence,

$$P(A \text{ or } B) = P(A) + P(B) - P(A \text{ and } B) = \frac{4}{52} + \frac{13}{52} - \frac{1}{52} = \frac{16}{52} = \frac{4}{13}.$$

EXAMPLE

A die is rolled. Find the probability of getting an odd number or a number less than 5.

 SOLUTION

Let A = an odd number; then $P(A) = \frac{3}{6}$ since there are three odd numbers—namely 1, 3, and 5. Let B = a number less than 5; then $P(B) = \frac{4}{6}$ since there are four numbers less than 5—namely 1, 2, 3, and 4. Let $P(A \text{ and } B) = \frac{2}{6}$ since there are two odd numbers less than 5—1 and 3. Hence,

$$P(A \text{ or } B) = P(A) + P(B) - P(A \text{ and } B) = \frac{3}{6} + \frac{4}{6} - \frac{2}{6} = \frac{5}{6}.$$

The results to both previous exercises can be verified by using sample spaces and classical probability.

EXAMPLE

Two dice are rolled; find the probability of getting doubles or a sum of 6.

SOLUTION

Let A = getting doubles; then $P(A) = \frac{6}{36}$ since there are six ways to get doubles and let B = getting a sum of 6. Then $P(B) = \frac{5}{36}$ since there are five ways to get a sum of 6—(5, 1), (4, 2), (3, 3), (2, 4), and (1, 5). Let A and B = the number of ways to get a double and a sum of 6. There is only one way for this event to occur—namely (3, 3); then $P(A \text{ and } B) = \frac{1}{36}$. Hence,

$$P(A \text{ or } B) = P(A) + P(B) - P(A \text{ and } B) = \frac{6}{36} + \frac{5}{36} - \frac{1}{36} = \frac{10}{36} = \frac{5}{18}.$$

 **EXAMPLE**

At a sales meeting, there were 15 men and 24 women. Eight of the men and 10 of the women were selected to sell a new product. If a salesperson were selected at random, find the probability that the salesperson was a female or a person who was selected to sell the new product.

 **SOLUTION**

There were 15 + 24 = 39 people at the meeting. Since there are 24 women, P(women) = $\frac{24}{39}$, and since 8 + 10 = 18 of the salespersons were selected to sell the new product, P(person selected to sell new product) = $\frac{18}{39}$. The number of sales people who are female and who were selected to sell the new product is 10. Hence,

$$P(\text{female or sales person}) = P(\text{female}) + P(\text{sales person})$$
$$- P(\text{female and sales person}) = \frac{24}{39} + \frac{18}{39} - \frac{10}{39} = \frac{32}{39}.$$

EXAMPLE

The probability that a student owns an iPod is 0.47, and the probability that a student owns a cell phone is 0.87. If the probability that a student owns both an iPod and a cell phone is 0.39, find the probability that the student owns an iPod or a cell phone.

SOLUTION

Since P(iPod) = 0.47, P(cell phone) = 0.87, and P(iPod and cell phone) = 0.39, P(iPod or cell phone) = 0.47 + 0.87 − 0.39 = 0.95.

The key word for addition is "or," and it means that one event or the other occurs. If the events are not mutually exclusive, the probability of the outcomes that the two events have in common must be subtracted from the sum of the probabilities of the two events. For the mathematical purist, only one addition rule is necessary, and that is

$$P(A \text{ or } B) = P(A) + P(B) - P(A \text{ and } B)$$

The reason is that when the events are mutually exclusive, $P(A \text{ and } B)$ is equal to zero because mutually exclusive events have no outcomes in common.

Still Struggling

When you are using the addition rule, avoid counting things twice when the events are not mutually exclusive.

 PRACTICE

1. Two dice are rolled. Find the probability that a number on one die is a 3 or the sum of the spots is 7.

2. When a card is selected at random from a 52 card deck, find the probability that the card is a face card or a club.

3. In a psychology class, there are 16 sophomores and 20 juniors. Seven of the sophomores are males and 11 of the juniors are males. If a student is selected at random, find the probability that the student is

   a. A junior or a female

   b. A sophomore or a male

   c. A junior

4. A die is rolled. Find the probability that the result is a number divisible by 3 or a number less than 4.

5. A coin is tossed and a die is rolled. Find the probability that the coin falls heads up or that there is a 4 on the die.

ANSWERS

1. $P(3 \text{ or a sum of } 7) = P(3) + P(\text{sum of } 7) - P(3 \text{ and a sum of } 7) =$ $\frac{11}{36} + \frac{6}{36} - \frac{2}{36} = \frac{15}{36} = \frac{5}{12}$

2. $P(\text{face card or club}) = P(\text{face card}) + P(\text{club}) - P(\text{face card and club}) =$ $\frac{12}{52} + \frac{13}{52} - \frac{3}{52} = \frac{22}{52} = \frac{11}{26}$

3. a. $P(\text{junior or female}) = P(\text{junior}) + P(\text{female}) - P(\text{junior and female}) =$ $\frac{20}{36} + \frac{18}{36} - \frac{9}{36} = \frac{29}{36}$

   b. $P(\text{sophomore or male}) = P(\text{sophomore}) + P(\text{male}) - P(\text{sophomore and male}) = \frac{16}{36} + \frac{18}{36} - \frac{7}{36} = \frac{27}{36} = \frac{3}{4}$

   c. $P(\text{junior}) = \frac{20}{36} = \frac{5}{9}$

4. $P(\text{number divisible by 3 or less than 4}) = P(\text{number divisible by 3}) + P(\text{less than 4}) - P(\text{number divisible by 3 and less than 4}) = \frac{2}{6} + \frac{3}{6} - \frac{1}{6} = \frac{4}{6} = \frac{2}{3}$

5. $P(\text{heads or 4}) = P(\text{heads}) + P(4) - P(\text{heads and 4}) = \frac{1}{2} + \frac{1}{6} - \frac{1}{12} = \frac{6}{12} + \frac{2}{12} - \frac{1}{12} = \frac{7}{12}$

PROBABILITY SIDELIGHT: Win a Million or Be Struck by Lightning?

Do you think you are more likely to win a large amount of money in a lottery and become a millionaire or are you more likely to be struck by lightning?

Consider each probability. In a recent article, researchers estimated that the chance of winning a million or more dollars is about one in 2 million. In a recent Pennsylvania State Lottery, the chances of winning a million dollars were 1 in 9.6 million. The chances of winning a $10 million prize in Publisher's Clearinghouse Sweepstakes were 1 in 2 million. Now the chances of being struck by lightning are about 1 in 600,000. Here a person is at least three times more likely to be struck by lightning than to win a million dollars!

But wait a minute! Statisticians are critical of these types of comparisons since winning the lottery is a random occurrence. But being struck by lightning depends on several factors. For example, if a person lives in a region where there are a lot of thunderstorms, his or her chances of being struck increase. Also, where a person is during a thunderstorm influences his or her chances of being struck by lightning. If the person is in a safe place such as inside a building or in an automobile, the probability of being struck is relatively small compared to a person standing out in a field or on a golf course during a thunderstorm.

So be wary of such comparisons. As the old saying goes, you cannot compare apples and oranges.

Summary

Many times in probability, it is necessary to find the probability of two or more events occurring. In these cases, the addition rules can be used. When the events are mutually exclusive, Addition Rule I is used, and when the events are not mutually exclusive, Addition Rule II is used. If the events are mutually exclusive, they have no outcomes in common. When the two events are not mutually exclusive, they have some common outcomes. The key word in these problems is "or," and it means to add.

QUIZ

1. **Which of the two events are mutually exclusive?**
    A. Drawing a card from a deck and getting a king or a club
    B. Rolling a die and getting an even number or a 6
    C. Tossing two coins and getting two heads or two tails
    D. Rolling two dice and getting doubles or getting a sum of 8

2. **Which of the two events are not mutually exclusive?**
    A. Rolling a die and getting a 6 or a 3
    B. Drawing a card from a deck and getting a club or an ace
    C. Tossing a coin and getting a head or a tail
    D. Tossing a coin and rolling a die and getting a head and an odd number

3. **In a box there are five dimes, two quarters, and three pennies. If a coin is selected at random, what is the probability that it is a quarter or a penny?**
    A. $\frac{2}{5}$
    B. $\frac{1}{3}$
    C. $\frac{9}{10}$
    D. $\frac{1}{2}$

4. **A storeowner plans to have his annual "Going Out of Business Sale." If each month has an equal chance of being selected, find the probability that the sale will be in a month that begins with the letter J or A.**
    A. $\frac{1}{4}$
    B. $\frac{1}{6}$
    C. $\frac{1}{3}$
    D. $\frac{5}{12}$

5. **A card is selected from a deck of 52 cards. Find the probability that it is a red king or a black queen.**
    A. $\frac{2}{13}$
    B. $\frac{1}{13}$
    C. $\frac{5}{13}$
    D. $\frac{8}{13}$

6. **When a single die is rolled, what is the probability of getting a prime number (2, 3, or 5) or a number less than 4?**
    A. $\frac{5}{6}$
    B. $\frac{2}{3}$
    C. $\frac{1}{2}$
    D. $\frac{1}{6}$

7. A card is selected from a deck. Find the probability that it is a 3 or a heart.

   A. $\frac{17}{52}$

   B. $\frac{4}{13}$

   C. $\frac{11}{26}$

   D. $\frac{13}{52}$

8. A single card is selected from a deck. Find the probability that it is a king or a red card.

   A. $\frac{11}{26}$

   B. $\frac{7}{13}$

   C. $\frac{1}{13}$

   D. $\frac{15}{26}$

9. At a high school with 200 students, 32 play soccer, 18 play basketball, and 8 play both sports. If a student is selected at random, find the probability that a student plays soccer or basketball.

   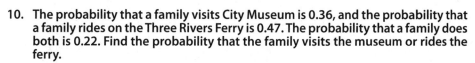

   A. $\frac{21}{100}$

   B. $\frac{1}{4}$

   C. $\frac{4}{25}$

   D. $\frac{1}{5}$

10. The probability that a family visits City Museum is 0.36, and the probability that a family rides on the Three Rivers Ferry is 0.47. The probability that a family does both is 0.22. Find the probability that the family visits the museum or rides the ferry.

    A. 0.83

    B. 0.61

    C. 0.58

    D. 0.69

chapter 4

The Multiplication Rules

The previous chapter showed how the addition rules could be used to solve problems in probability. This chapter will show you how to use the multiplication rules to solve many problems in probability. In addition, the concept of independent and dependent events will be explained.

CHAPTER OBJECTIVES

In this chapter, you will learn

- The difference between independent and dependent events
- How to find the probability of two events using the multiplication rules
- How to find the conditional probability of an event

Independent and Dependent Events

The multiplication rules can be used to find the probability of two or more events that occur in sequence. For example, we can find the probability of selecting three jacks from a deck of cards on three sequential draws. Before explaining the rules, it is necessary to differentiate between **independent** and **dependent** events.

Two events, A and B, are said to be **independent** if the fact that event A occurred does not affect the probability that event B occurs. For example, if a coin is tossed and then a die is rolled, the outcome of the coin in no way affects the outcome, or changes the probability of the outcome, of the die. Another example would be selecting a card from a deck, replacing it, and then selecting a second card from the same deck; the outcome of the first card has no effect on the probability or the outcome of the second card selected.

On the other hand, when the occurrence of the first event in some way changes the probability of the occurrence of the second event, the two events are said to be **dependent**. For example, suppose a card is selected from a deck and **not** replaced, and a second card is selected. In this case, the probability of selecting any specific card on the first draw is $\frac{1}{52}$, but since this card is not replaced, the probability of selecting any other specific card is $\frac{1}{51}$ since there are only 51 cards left.

Another example would be parking in a no-parking zone and getting a parking ticket. Again, if you are legally parked, the chances of getting a parking ticket are pretty close to 0 (as long as the meter does not run out). However, if you are illegally parked, your chance of getting a parking ticket dramatically increases.

 **PRACTICE**

Determine whether the two events are independent or dependent.

1. Tossing a coin and selecting a card from a deck

2. Driving on ice and having an accident

3. Drawing a ball from an urn, not replacing it, and then drawing a second ball

4. Having a high I.Q. and having a large hat size

5. Tossing one coin and then tossing a second coin

**ANSWERS**

1. **Independent. Tossing a coin has no effect on drawing a card.**

2. **Dependent. In most cases, driving on ice will increase the probability of having an accident.**

3. **Dependent. Since the first ball is not replaced before the second ball is selected, it will change the probability of a specific second ball being selected.**

4. **Independent. To the best of the author's knowledge, no studies have been done showing any relationship between hat size and I.Q.**

5. **Independent. The outcome of the first does not influence the outcome of the second coin.**

Multiplication Rule I

Before explaining the first multiplication rule, consider the example of tossing two coins. The sample space is HH, HT, TH, TT. From classical probability theory, it can be determined that the probability of getting two heads is $\frac{1}{4}$ since there is only one way to get two heads and there are four outcomes in the sample space. However, there is another way to determine the probability of getting two heads. In this case, the probability of getting a head on the first toss is $\frac{1}{2}$, and since the events are independent, the probability of getting a head on the second toss is also $\frac{1}{2}$. So the probability of getting two heads can be determined by multiplying $\frac{1}{2} \cdot \frac{1}{2} = \frac{1}{4}$. This example illustrates the first multiplication rule.

Multiplication Rule I: For two independent events A and B, $P(A \text{ and } B) = P(A) \cdot P(B)$.

In other words, when two independent events occur in sequence, the probability that both events will occur can be found by multiplying the probabilities of each individual event.

Still Struggling

In this chapter, the word "and" means that the two events occur in sequence. Note: In the previous chapters, the word "and" meant "at the same time" since the experiment was done only one time.

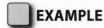 **EXAMPLE**

A coin is tossed and a die is rolled. Find the probability of getting a head on the coin and a 4 on the die.

 SOLUTION

Since $P(\text{head}) = \frac{1}{2}$ and $P(4) = \frac{1}{6}$, $P(\text{head and } 4) = P(\text{head}) \cdot P(4) = \frac{1}{2} \cdot \frac{1}{6} = \frac{1}{12}$. Note that the events are independent.

The previous example can also be solved using classical probability. Recall that the sample space for tossing a coin and rolling a die is

$$H1, H2, H3, H4, H5, H6$$
$$T1, T2, T3, T4, T5, T6$$

Notice that there are 12 outcomes in the sample space and only one outcome is a head and a 4; hence, $P(\text{head and } 4) = \frac{1}{12}$.

EXAMPLE

An urn contains three red balls, four green balls, and two blue balls. A ball is selected at random and its color is noted. Then it is replaced and another ball is selected and its color is noted. Find the probability of each of these:

  a. Selecting two green balls

  b. Selecting a blue ball then a red ball

  c. Selecting a green ball then a red ball

SOLUTION

Since the first ball is being replaced before the second ball is selected, the events are independent.

  a. There are four green balls and a total of nine balls; therefore, the probability of selecting two green balls with replacement is

$$P(\text{green and green}) = P(\text{green}) \cdot P(\text{green})$$
$$= \frac{4}{9} \cdot \frac{4}{9} = \frac{16}{81}$$

  b. There are two blue balls and three red balls, so the probability of selecting a blue ball and then a red ball with replacement is

$$P(\text{blue and red}) = P(\text{blue}) \cdot P(\text{red})$$
$$= \frac{2}{\cancel{9}^{3}} \cdot \frac{\cancel{3}^{1}}{9} = \frac{2}{3 \cdot 9} = \frac{2}{27}$$

c. There are four green balls and three red balls, so the probability of selecting a green ball and then a red ball with replacement is

$$P(\text{green and red}) = P(\text{green}) \cdot P(\text{red})$$

$$= \frac{4}{\cancel{9}^{3}} \cdot \frac{\cancel{3}^{1}}{9} = \frac{4}{27}$$

The multiplication rule can be extended to three or more events that occur in sequence, as shown in the next example.

EXAMPLE _____

A coin is tossed three times. Find the probability of getting three tails.

SOLUTION _____

When a coin is tossed, the probability of getting a tail is $\frac{1}{2}$; hence, the probability of getting three tails is

$$P(3 \text{ tails}) = P(T) \cdot P(T) \cdot P(T) = \frac{1}{2} \cdot \frac{1}{2} \cdot \frac{1}{2} = \frac{1}{8}$$

Another situation that occurs in probability is when subjects are selected from a large population. Even though the subjects are not replaced, the probability changes only slightly, so the concern of replacement versus nonreplacement can be ignored. Consider the next example.

EXAMPLE _____

Approximately 9% of men have a type of color blindness that prevents them from distinguishing between red and green. If two men are selected at random from a large group of men, find the probability that both will have this type of color blindness.

SOLUTION _____

Since the group of men is very large, selecting one man does not radically change the probability that another man will be selected. Let C = red-green color blindness; then $P(C \text{ and } C) = P(C) \cdot P(C) = (0.09)(0.09) = 0.0081$.

PRACTICE

1. A coin is tossed five times. Find the probability of getting five heads.

2. A card is drawn from a deck then replaced, and a second card is drawn. Find the probability that two aces are selected.

3. If 12% of adults are left-handed, find the probability that if two adults are selected at random, both will be left-handed.

4. If two people are selected at random, find the probability that both were born on Wednesday.

5. A coin is tossed and a card is selected at random from a deck of 52 cards. Find the probability of getting a head and a diamond on the card.

✓ANSWERS

1. The probability of getting five heads is $\frac{1}{2} \cdot \frac{1}{2} \cdot \frac{1}{2} \cdot \frac{1}{2} \cdot \frac{1}{2} = \frac{1}{32}$.

2. The probability that two aces are selected is

$$P(\text{ace and ace}) = P(\text{ace}) \cdot P(\text{ace}) = \frac{\overset{1}{\cancel{4}}}{\underset{13}{\cancel{52}}} \cdot \frac{\overset{1}{\cancel{4}}}{\underset{13}{\cancel{52}}} = \frac{1}{169}$$

3. The probability of selecting two people who are left-handed is (0.12)(0.12) = 0.0144.

4. Each person has approximately one chance in seven of being born on Wednesday; hence, the probability that two people are born on Wednesday is $\frac{1}{7} \cdot \frac{1}{7} = \frac{1}{49}$.

5. The probability of getting a head is $\frac{1}{2}$, and the probability of getting a diamond is $\frac{13}{52} = \frac{1}{4}$; hence, the $P(\text{head and diamond}) = P(H) \cdot P(\text{diamond}) = \frac{1}{2} \cdot \frac{1}{4} = \frac{1}{8}$.

Multiplication Rule II

When two sequential events are dependent, a slight variation of the multiplication rule is used to find the probability of both events occurring. For example, when a card is selected from an ordinary deck of 52 cards, the probability of getting a specific card is $\frac{1}{52}$, but the probability of getting a specific card on the second draw is $\frac{1}{51}$ since 51 cards remain.

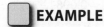**EXAMPLE**

Two cards are selected from a deck and the first card is **not** replaced. Find the probability of getting two queens.

SOLUTION

The probability of getting a queen on the first draw is $\frac{4}{52}$ and the probability of getting a queen on the second draw is $\frac{3}{51}$ since there are three queens and 51 cards left. Hence, the probability of getting two queens when the first card is not replaced is $\frac{\overset{1}{\cancel{4}}}{\underset{13}{\cancel{52}}} \cdot \frac{\overset{1}{\cancel{3}}}{\underset{17}{\cancel{51}}} = \frac{1}{221}$.

When the two events A and B are dependent, the probability that the second event B occurs after the first event A has already occurred is written as $P(B|A)$. This does not mean that B is divided by A; rather, it means and is read as "the probability that event B occurs *given* that event A has already occurred." $P(B|A)$ also means the **conditional** probability that event B occurs given event A has occurred. The second multiplication rule follows.

Multiplication Rule II: When two events are dependent, the probability of both events occurring is $P(A \text{ and } B) = P(A) \cdot P(B|A)$.

EXAMPLE

A box contains 18 CDs, four of which are defective. If two CDs are selected and tested, find the probability that both are defective.

SOLUTION

Since there are four defective CDs out of 18, the probability that the first CD is defective is $\frac{4}{18}$. Since the second CD is selected from the remaining 17 and there are three defective CDs left, the probability that it is defective is $\frac{3}{17}$. Hence, the probability that both CDs are defective is $P(D_1 \text{ and } D_2) = P(D_2|D_1) = \frac{4}{18} \cdot \frac{3}{17} = \frac{2}{51}$.

EXAMPLE

Two cards are drawn without replacement from a deck of 52 cards. Find the probability that they are both clubs.

SOLUTION

There are 52 cards and 13 clubs, so the probability that the first card is a club is $\frac{13}{52}$. There are 51 cards and 12 clubs left, so the probability that the second card is a club given the first card was a club and not replaced is $\frac{12}{51}$. Hence, $P(\text{club and club}) = P(\text{club}) \cdot P(\text{club}|\text{club}) = \frac{13}{52} \cdot \frac{12}{51} = \frac{1}{17}$.

This multiplication rule can be extended to include 3 or more events, as shown in the next example.

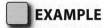**EXAMPLE**

A football team has 22 players. Eleven play offense and 11 play defense. If three players are selected at random without replacement, find the probability that all three will be offensive players.

**SOLUTION**

$$P(\text{offense and offense and offense}) = \frac{11}{22} \cdot \frac{10}{21} \cdot \frac{9}{20} = \frac{\cancel{11}^{1}}{\cancel{22}^{2}} \cdot \frac{\cancel{10}^{1}}{\cancel{21}^{7}} \cdot \frac{\cancel{9}^{3}}{\cancel{20}^{2}} = \frac{3}{28}.$$

Remember that the key word for the multiplication rule is *and*. It means to multiply.

When two events are dependent, the probability that the second event occurs must be adjusted for the occurrence of the first event. For the mathematical purist, only one multiplication rule is necessary for two events, and that is

$$P(A \text{ and } B) = P(A) \cdot P(B|A).$$

The reason is that when the events are independent, $P(B|A) = P(B)$, since the occurrence of the first event A has no effect on the occurrence of the second event B.

PRACTICE

1. **In a small classroom there are 10 boys and 12 girls. If two students are selected at random without replacement to pass out paper, find the probability that both would be girls.**

2. **If a person is dealt three cards from a deck of 52 cards, find the probability that all three cards are hearts.**

3. **A carton of light bulbs contains 15 bulbs, of which four are defective. If three bulbs are selected at random without replacement and tested, find the probability that they are all good.**

4. **In a mathematics class of 24 students, eight received an A for the course. If three students are randomly selected without replacement, find the probability that all of them received an A.**

5. A box contained eight blue marbles and six red marbles. If three marbles are selected at random without replacement, find the probability that the first marble is blue and the other two marbles are red.

ANSWERS

1. $P(\text{girl and girl}) = \frac{12}{22} \cdot \frac{11}{21} = \frac{2}{7}$

2. $P(3 \text{ hearts}) = \frac{13}{52} \cdot \frac{12}{51} \cdot \frac{11}{50} = \frac{11}{850}$

3. There are four defective bulbs and 11 good ones. So $P(3 \text{ good bulbs}) =$ $\frac{11}{15} \cdot \frac{10}{14} \cdot \frac{9}{13} = \frac{33}{91}$

4. $P(3\text{As}) = \frac{8}{24} \cdot \frac{7}{23} \cdot \frac{6}{22} = \frac{7}{253}$

5. $P(\text{blue and red and red}) = \frac{8}{14} \cdot \frac{6}{13} \cdot \frac{5}{12} = \frac{10}{91}$

Conditional Probability

Previously in this chapter conditional probability was used to find the probability of sequential events occurring when they were dependent. Recall that $P(B|A)$ means the probability of event B occurring given that event A has already occurred. Another situation where conditional probability can be used is when additional information about an event is known. Sometimes it might be known that some outcomes in the sample space have occurred or that some outcomes cannot occur. When conditions are imposed or known on events, there is a possibility that the probability of the certain event occurring is changed. For example, suppose you want to determine the probability that a house will be destroyed by a hurricane. If you used all houses in the United States as the sample space, the probability would be very small. However, if you used only the houses in the states that border the Atlantic Ocean as the sample space, the probability would be much higher. Consider the following examples.

EXAMPLE

A card is selected from a deck. Find the probability that it is an ace given that it is a red card.

**SOLUTION**

If it is known that the card is a red card, the sample space is reduced to the 26 red cards. Hence, the probability of getting an ace is $\frac{2}{26} = \frac{1}{13}$ since there are two red aces.

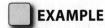**EXAMPLE**

Two dice are rolled. Find the probability of getting a sum of 5 if it is known that the sum of the spots on the dice was odd.

**SOLUTION**

There are four ways to get a sum of 5. They are (1, 4), (2, 3), (3, 2), (4, 1), and there are 18 ways to get a sum that is odd; hence, P(sum of 5|sum is odd) $= \frac{4}{18} = \frac{2}{9}$.

The two previous examples of conditional probability were solved using classical probability and reduced sample spaces; however, they can be solved by using the following formula for conditional probability.

The conditional probability of two events A and B is $P(A|B) = \frac{P(A \text{ and } B)}{P(B)}$.

$P(A \text{ and } B)$ means the probability of the outcomes that events A and B have in common. The two previous examples will now be solved using the formula for conditional probability.

Still Struggling

This formula can be derived from the previous formula $P(A \text{ and } B) = P(A) \cdot P(B|A)$ by interchanging A and B and solving the equation for $P(A|B)$ since $P(A \text{ and } B) = P(B \text{ and } A)$.

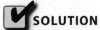**EXAMPLE**

A card is selected from a deck. Find the probability that it is an ace given that it is a red card.

SOLUTION

$P(A \text{ and } B)$ means getting an ace and a red card. Hence, $P(A \text{ and } B) = \frac{2}{52}$ since there are two red aces. $P(B)$ is the outcome of getting a red card, which is $\frac{1}{2}$.

$$P(A|B) = \frac{P(A \text{ and } B)}{P(B)}$$

$$= \frac{\dfrac{2}{52}}{\dfrac{1}{2}}$$

$$= \frac{2}{52} \div \frac{1}{2} = \frac{2}{52} \cdot \frac{2}{1} = \frac{4}{52} = \frac{1}{13}$$

Notice that the answer is the same as the answer obtained when classical probability was used.

EXAMPLE

Two dice are rolled. Find the probability of getting a sum of 5 if it is known that the sum of the spots on the dice is odd.

SOLUTION

$P(A \text{ and } B)$ means the probability of getting a sum of 5 and a sum that is odd. Hence, $P(A \text{ and } B) = \frac{4}{36}$. $P(B)$ means getting a sum that is odd and is $\frac{1}{2}$.

$$P(A|B) = \frac{P(A \text{ and } B)}{P(B)} = \frac{\dfrac{4}{36}}{\dfrac{1}{2}}$$

$$= \frac{4}{36} \div \frac{1}{2} = \frac{\overset{1}{\cancel{4}}}{\underset{9}{\cancel{36}}} \cdot \frac{2}{1} = \frac{2}{9}$$

EXAMPLE

Three coins are tossed. Find the probability of getting three heads if it is known that at least one head appeared on one of the coins.

SOLUTION

There is one way to get three heads, so $P(3 \text{ heads})$ is $\frac{1}{8}$. There are seven ways that at least one head can appear on the three coins. Thus,

Let A = Three heads appear

Let B = At least one head appears

Then $\qquad P(A|B) = \dfrac{P(A \text{ and } B)}{P(B)}$

$$= \dfrac{P(3 \text{ heads and at least 1 head})}{P(\text{at least 1 head})}$$

$$= \dfrac{P(3 \text{ heads})}{P(\text{at least 1 head})}$$

$$= \dfrac{\dfrac{1}{8}}{\dfrac{7}{8}} = \dfrac{1}{\cancel{8}^{1}} \cdot \dfrac{\cancel{8}^{1}}{7} = \dfrac{1}{7}$$

EXAMPLE

In a fast food restaurant, it was found that 40% of the customers ordered a cheeseburger and fries, and 64% of the customers ordered a cheeseburger. Find the probability that a person ordered fries given that the person ordered a cheeseburger.

SOLUTION

$$P(\text{cheeseburger and fries}) = 0.40,\ P(\text{cheeseburger}) = 0.64$$

$$P(\text{fries}|\text{cheeseburger}) = \dfrac{P(\text{fries and cheeseburger})}{P(\text{cheeseburger})} = \dfrac{0.40}{0.64} = \dfrac{5}{8} = 0.625$$

PRACTICE

1. When two dice are rolled, find the probability of getting doubles given that the sum of the spots is 10.

2. Two coins are tossed. Find the probability of getting two heads if it is known that one of the coins is a head.

3. A card is selected from a deck. Find the probability that it is a club given that it is a black card.

4. At an automobile dealer, it is known that 32% of the cars sold are white, 2-door models. If 40% of the cars sold are 2-door models, find the probability that if a car is sold, it is white given that it is a 2-door model.

5. Three dice are rolled. Find the probability of getting exactly one 3 if it is known that the sum of the spots on the three dice was 5.

ANSWERS

1. There are three ways to get a sum of 10. They are (4, 6), (5, 5), and (6, 4), and one way to get a double when a sum of 10 has occurred. Hence, P(doubles|a sum of 10) =

$$\frac{P(A \text{ and } B)}{P(B)} = \frac{\frac{1}{36}}{\frac{3}{36}} = \frac{1}{36} \div \frac{3}{36} = \frac{1}{\cancel{36}^{1}} \cdot \frac{\cancel{36}^{1}}{3} = \frac{1}{3}.$$

2. There are three ways to get at least one head. They are HT, TH, and HH. (Note: HH is included since the problem does not say exactly one head.) There is one way to get 2 heads:

$$P = \frac{P(2 \text{ heads})}{P(\text{at least 1 coin is a head})} = \frac{1}{3}$$

   The problem can also be done as follows:

$$P = \frac{P(2 \text{ heads})}{P(\text{at least 1 coin is a head})}$$

$$= \frac{P(2 \text{ heads and 1 coin is a head})}{P(\text{at least 1 coin is a head})}$$

$$= \frac{P(2 \text{ heads})}{P(\text{at least 1 coin is a head})}$$

$$= \frac{\frac{1}{4}}{\frac{3}{4}} = \frac{1}{4} \div \frac{3}{4} = \frac{1}{\cancel{4}^{1}} \cdot \frac{\cancel{4}^{1}}{3} = \frac{1}{3}.$$

3. The probability of a black card is $\frac{26}{52} = \frac{1}{2}$, and the probability of a club is $\frac{13}{52} = \frac{1}{4}$. Hence, P(club|a black card) $= \frac{\frac{1}{4}}{\frac{1}{2}} = \frac{1}{4} \div \frac{1}{2} = \frac{1}{\cancel{4}^{2}} \cdot \frac{\cancel{2}^{1}}{1} = \frac{1}{2}$, since P(club and black) reduces to P(club) $= \frac{1}{4}$.

4. The probability of a 2-door white automobile is 0.32, and the probability of a 2-door car is 0.40. Hence,

$$P(\text{white}|\text{2-door}) = \frac{P(\text{white and 2-door})}{P(\text{2-door})} = \frac{0.32}{0.40} = \frac{4}{5} = 0.80.$$

5. There are six ways to get a sum of 5. They are (1, 1, 3), (1, 3, 1), (3, 1, 1), (1, 2, 2), (2, 1, 2), and (2, 2, 1), and three ways to get exactly one 3, so P(exactly one 3|sum of spots is 5) $= \frac{3}{6} = \frac{1}{2}$.

PROBABILITY SIDELIGHT: The Law of Averages

Suppose I asked you that if you tossed a coin nine times and got nine heads, what would you bet that you would get on the 10th toss, heads or tails? Most people would bet on a tail. When asked why they would select a tail, they would probably respond that a tail was "due" according to the "law of averages." In reality, however, the probability of getting a head on the 10th toss is $\frac{1}{2}$, and the probability of getting a tail on the 10th toss is also $\frac{1}{2}$, so it doesn't really matter since the probabilities are the same. A coin is an inanimate object. It does not have a memory. It doesn't know that in the long run, the number of heads and the number of tails should balance out. So does that make the law of averages wrong? No. You see, there's a big difference between asking the question, "What is the probability of getting 10 heads if I toss a coin 10 times?" and "If I get 9 heads in a row, what is the probability of getting a head on the 10th toss?" The answer to the first question is $\frac{1}{2^{10}} = \frac{1}{1024}$, that is, about 1 chance in 1000, and the answer to the second question is $\frac{1}{2}$.

This reasoning can be applied to many situations. For example, suppose that a prize is offered for tossing a coin and getting 10 heads in a row. If you played the game, you would have only one chance in 1024 of winning, but if 1024 people played the game, there is a pretty good chance that somebody would win the prize. If 2028 people played the game, there would be a good chance that two people might win. So what does this mean? It means that the probability of winning big in a lottery or on a slot machine is very small, but since there are many, many people playing, somebody will probably win; however, your chances of winning big are very small.

A similar situation occurs when couples have children. Suppose a husband and wife have four boys and would like to have a girl. The incorrect reasoning is that the chance of having a family of five boys is $\frac{1}{32}$, so it is more likely that the next child will be a girl. However, after each child is born, the probability that the next child is a girl (or a boy for that matter) is about $\frac{1}{2}$. The law of averages is not appropriate here.

My wife's aunt had seven girls before the first boy was born. Also, in the Life Science Library's book entitled *Mathematics,* there is a photograph of the Landon family of Harrison, Tennessee, that shows Mr. and Mrs. Emery Landon and their 13 boys!

Another area where people incorrectly apply the law of averages is in attempting to apply a betting system to gambling games. One such system is doubling your bet when you lose. Consider a game where a coin is tossed. If it lands heads, you win what you bet; if it lands tails, you lose. Now if you bet $1 on the first toss and get a head, you win $1; if you get tails, you lose $1 and bet $2 on the next toss. If you win, you are $1 ahead because you lost $1 on the first bet but won $2 on the second bet. If you get a tail on the second toss, you bet $4 on the third toss.

If you win, you start over with a $1 bet, but if you lose, you bet $8 on the next toss. With this system, you win every time you get a head. Sounds pretty good, doesn't it?

This strategy won't work because if you play long enough, you will eventually run out of money since if you get a series of tails, you must increase your bet substantially each time. So if you lose five times in a row, you have lost $1 + $2 + $4 + $8 + $16 or $31, and your next bet has to be $32. So you are betting $63 to win $1. Runs do occur and when they do, hope that they are in your favor.

Now let's look at some unusual so-called "runs."

In 1950, a person won 28 straight times playing the game of craps (dice) at the Desert Inn in Las Vegas. He lost on the 29th roll. He did not win big, though, because after each win he stuffed some bills in his pocket. The event took about 1 hour and 20 minutes.

In 1959 in a casino in Puerto Rico at a roulette game, the number 10 occurred six times in succession. There are 38 numbers on a roulette wheel.

At a casino in New York in 1943, the color red occurred in a roulette game 32 times in a row, and at a casino in Monte Carlo, an even number occurred in a roulette game 28 times in a row.

These incidents have been reported in two books, one entitled *Scarne's Complete Guide to Gambling* and the other entitled *Lady Luck* by Warren Weaver.

So what can be concluded? First, rare events (events with a small probability of occurring) can and do occur. Second, the more people who play a game, the more likely someone will win. Finally, the law of averages applies when there is a large number of independent outcomes in which the probability of each outcome occurring does not change.

Summary

When two events occur in sequence, the probability that both events occur can be found by using one of the multiplication rules. When the events are independent, the probability that the first event occurs does not affect or change the probability of the second event occurring; in this case, Multiplication Rule I is used. When the two events are dependent, the probability of the second event occurring is changed after the first event occurs. If the events are dependent, Multiplication Rule II is used. The key word for using the multiplication rule is "and." Conditional probability is used when additional information is known about the probability of an event.

QUIZ

1. Which of the following events are dependent?
   A. Tossing a coin and selecting a card from a deck
   B. Tossing a coin then tossing a second coin
   C. Running a race and getting tired
   D. Drawing a card from a deck and replacing it, then drawing a second card

2. Three dice are rolled. What is the probability of getting three 6s?

   A. $\frac{1}{6}$

   B. $\frac{1}{216}$

   C. $\frac{1}{18}$

   D. $\frac{5}{36}$

3. What is the probability of selecting four red cards from a deck of 52 cards if each card is replaced before the next one is selected?

   A. $\frac{1}{13}$

   B. $\frac{1}{256}$

   C. $\frac{1}{52}$

   D. $\frac{1}{16}$

4. A die is rolled twice. What is the probability of getting two even numbers?

   A. $\frac{1}{4}$

   B. $\frac{1}{6}$

   C. $\frac{1}{32}$

   D. $\frac{1}{8}$

5. A coin is tossed six times; what is the probability of getting six tails?

   A. $\frac{1}{2}$

   B. $\frac{1}{4}$

   C. $\frac{1}{12}$

   D. $\frac{1}{64}$

6. If 3% of ties sold in the United States are bow ties, what is the probability of randomly selecting three ties that are sold and all three are bow ties?

   A. 0.027

   B. 0.09

   C. 0.9

   D. 0.000027

7. If three people are randomly selected, find the probability that they will all have birthdays on the same day of the year. (1 year = 365 days.)

   A. $\frac{1}{133,225}$

   B. $\frac{3}{365}$

   C. $\frac{1}{1095}$

   D. $\frac{1}{48,627,125}$

8. A box contains four $1 bills and six $5 bills. If three bills are selected at random without replacement, find the probability that all three are $5 bills.

   A. $\frac{27}{125}$

   B. $\frac{1}{4}$

   C. $\frac{1}{6}$

   D. $\frac{3}{5}$

9. The numbers 1 to 15 are placed in a hat, and a number is selected. What is the probability that the number is 4 given that it is known to be an even number?

   A. $\frac{1}{15}$

   B. $\frac{1}{7}$

   C. $\frac{1}{2}$

   D. $\frac{1}{8}$

10. Two dice are tossed; what is the probability that the numbers were the same on both dice if it is known that the sum of the spots is 4?

    A. $\frac{1}{6}$

    B. $\frac{4}{5}$

    C. $\frac{1}{3}$

    D. $\frac{1}{36}$

chapter 5

Odds and Expectation

In this chapter, you will learn about two concepts that are often used in conjunction with probability. They are **odds** and **expectation**. Odds are used most often in gambling games at casinos, racetracks, in sports betting, and lotteries. Odds make it easier than probabilities to determine payoffs.

Mathematical expectation can be thought of more or less as an average over the long run. In other words, if you would perform a probability experiment many times, the expectation would be an average of the outcomes. Also, expectation can be used to determine the average payoff per game in a gambling game.

CHAPTER OBJECTIVES

In this chapter, you will learn

- How to find the odds for and against an outcome for a probability event
- How to find the expected value of a probability experiment

Odds

Odds are used by casinos, racetracks, and other gambling establishments to determine payoffs when bets are made. For example, at a race, the odds that a horse wins the race may be 4 to 1. In this case, if you bet $1 and the horse wins, you get $4. If you bet $2 and the horse wins, you get $8, and so on.

Odds are computed from probabilities. For example, suppose you roll a die and if you roll a 3, you win. If you roll any other number, you lose. Furthermore, if you bet $1 and win, what would the payoff be if you win? In this case, there are six outcomes, and you have one chance (outcome) of winning, so the probability that you win is $\frac{1}{6}$. That means *on average* you win once in every six rolls. So if you lose on the first five rolls and win on the sixth, you have lost $5; therefore, you should get $5 if you win on the sixth roll. So if you bet $1 and win $5, the odds are 1 to 5. Of course, there is no guarantee that you will win on the sixth roll. You may win on the first roll or any roll, but *on average* for every six rolls, you will win one time over the long run.

In gambling games, the odds are expressed in reverse order. For example, if there is one chance in six that you will win, the odds are 1 to 5, but in general, the odds would be given as 5 to 1. In gambling, the house (the people running the game) will offer lower odds, say 4 to 1, in order to make a profit. In this case, then, the player wins on average one time in every six rolls and spends on average $5, but when the player wins, he gets only $4. So the house wins on average $1 for every six rolls of the player.

Odds can be expressed as a fraction, $\frac{1}{5}$, or as a ratio, 1:5. If the odds of winning the game are 1:5, then the odds of losing are 5:1. The odds of winning the game can also be called the odds "in favor" of the event occurring. The odds of losing can also be called "the odds against" the event occurring.

The formulas for odds are

$$\text{odds in favor} = \frac{P(E)}{1 - P(E)}$$

$$\text{odds against} = \frac{P(\overline{E})}{1 - P(\overline{E})}$$

where $P(E)$ is the probability that the event E occurs and $P(\overline{E})$ is the probability that the event does not occur.

 **EXAMPLE** _____

One die is rolled; what are the odds in favor of getting a 5 or 6?

 SOLUTION _____

When a die is rolled, there are six possible outcomes, and there are two ways to win. That is, getting a 5 or 6.

$$\text{Odds in favor of getting a 5 or 6} = \frac{P(E)}{1-P(E)} = \frac{\frac{2}{6}}{1-\frac{2}{6}} = \frac{\frac{2}{6}}{\frac{4}{6}} = \frac{2}{6} \div \frac{4}{6} = \frac{\cancel{2}^1}{\cancel{6}^1} \cdot \frac{\cancel{6}^1}{\cancel{4}^2} = \frac{1}{2}.$$

The odds are 1:2.

_____

 **EXAMPLE** _____

Two dice are rolled; find the odds against getting a sum of 8.

 SOLUTION _____

There are 36 outcomes in the sample space and five ways to get a sum of 8.

$$P(\text{sum of 8}) = \frac{5}{36}, \quad P(\overline{E}) = 1 - \frac{5}{36} = \frac{31}{36}.$$

Hence,

$$\text{Odds of not getting a sum of 8} = \frac{P(\overline{E})}{1-P(\overline{E})} = \frac{\frac{31}{36}}{\frac{5}{36}} = \frac{31}{36} \div \frac{5}{36} = \frac{31}{\cancel{36}^1} \cdot \frac{\cancel{36}^1}{5} = \frac{31}{5}.$$

The odds are 31:5.

_____

If the odds in favor of an event occurring are $A:B$, then the odds against the event occurring are $B:A$. For example, if the odds are 1:15 that an event will occur, then the odds against the event occurring are 15:1.

Odds can also be expressed as

$$\text{Odds in favor} = \frac{\text{number of outcomes in favor of the event}}{\text{number of outcomes not in favor of the event}}$$

For example, when a die is rolled, the odds in favor of getting a 5 or 6 are 1:2, as previously shown. Notice that there are two ways of getting a 5 or 6 and four ways of not getting a 5 or 6; hence, the odds are 2:4 or 1:2.

When the probability of an event occurring is $\frac{1}{2}$, then the odds are 1:1. In the realm of gambling, we say the odds are "even" and the chance of the event is "50-50." The game is said to be fair. Odds can be other numbers, such as 2:5, 7:4, etc.

 PRACTICE

1. When a single card is drawn from a deck of 52 cards, find the odds against getting a queen.

2. When three coins are tossed, find the odds in favor of getting exactly two tails.

3. When two dice are rolled, find the odds in favor of getting a sum of 7.

4. When two dice are rolled, find the odds in favor of getting a sum of 11.

5. When a single die is rolled, find the odds in favor of getting a number greater than one.

 **ANSWERS**

1. There are four queens in 52 cards; hence, $P(Q) = \frac{4}{52} = \frac{1}{13}$. $P(\overline{Q}) = 1 - \frac{1}{13} = \frac{12}{13}$.

   The odds against getting a queen are

   $$\frac{\frac{12}{13}}{\frac{1}{13}} = \frac{12}{13} \div \frac{1}{13} = \frac{12}{13^1} \cdot \frac{13^1}{1} = \frac{12}{1}$$

   The odds are 12:1.

2. When three coins are tossed, there are three ways to get two tails. There are eight outcomes in the sample space. The odds in favor of two tails is

   $$\frac{\frac{3}{8}}{1 - \frac{3}{8}} = \frac{\frac{3}{8}}{\frac{5}{8}} = \frac{3}{8} \div \frac{5}{8} = \frac{3}{8^1} \cdot \frac{8^1}{5} = \frac{3}{5}$$

   The odds are 3:5.

3. There are six ways to get a sum of 7 and 36 outcomes in the sample space. Hence, $P(\text{sum of 7}) = \frac{6}{36} = \frac{1}{6}$ and $P(\text{not getting a sum of 7}) = 1 - \frac{1}{6} = \frac{5}{6}$. The odds in favor of getting a sum of 7 are

$$\frac{\frac{1}{6}}{\frac{5}{6}} = \frac{1}{6} \div \frac{5}{6} = \frac{1}{\cancel{6}^1} \cdot \frac{\cancel{6}^1}{5} = \frac{1}{5}$$

The odds are 1:5.

4. There are two ways to get a sum of 11, and they are (5, 6) and (6, 5). There are 36 outcomes in the sample space. Hence, $P(\text{sum of 11}) = \frac{2}{36} = \frac{1}{18}$. The odds in favor are

$$\frac{\frac{1}{18}}{1 - \frac{1}{18}} = \frac{\frac{1}{18}}{\frac{17}{18}} = \frac{1}{18} \div \frac{17}{18} = \frac{1}{\cancel{18}^1} \cdot \frac{\cancel{18}^1}{17} = \frac{1}{17}$$

The odds are 1:17.

5. There are five ways out of six outcomes to get a number greater than one; hence, $P(\text{number} > 1)\ \frac{5}{6}$. The odds in favor of a number greater than one are

$$\frac{\frac{5}{6}}{1 - \frac{5}{6}} = \frac{\frac{5}{6}}{\frac{1}{6}} = \frac{5}{6} \div \frac{1}{6} = \frac{5}{\cancel{6}^1} \cdot \frac{\cancel{6}^1}{1} = \frac{5}{1}$$

The odds are 5:1.

Previously it was shown that given the probability of an event, the odds in favor of the event occurring or the odds against the event occurring can be found. The opposite is also true. If you know the odds in favor of an event occurring or the odds against an event occurring, you can find the probability of the event occurring. If the odds in favor of an event occurring are A:B, then the probability that the event will occur is $P(E) = \frac{A}{A+B}$.

If the odds against the event occurring are B:A, the probability that the event will not occur is $P(\bar{E}) = \frac{B}{B+A}$.

NOTE *Recall that $P(\overline{E})$ is the probability that the event will not occur or the probability of the complement of event E.*

Still Struggling

The formula $P(E) = \frac{A}{A+B}$ where $A:B$ are the odds that the event occurs can be derived from the formula $\frac{P(E)}{P(1-E)}$ and analogously for $P(\overline{E}) = \frac{B}{B+A}$.

EXAMPLE

If the odds that an event will occur are 3:8, find the probability that the event will occur.

SOLUTION

In this case, $A = 3$ and $B = 8$; hence, $P(E) = \frac{A}{A+B} = \frac{3}{3+8} = \frac{3}{11}$. Hence, the probability the event will occur is $\frac{3}{11}$.

EXAMPLE

If the odds in favor of an event are 4:9, find the probability that the event will **not** occur.

SOLUTION

In this case, $A = 4$ and $B = 9$; hence, the probability that the event will not occur is $P(\overline{E}) = \frac{B}{B+A} = \frac{9}{9+4} = \frac{9}{13}$.

PRACTICE

1. Find the probability that an event E will occur if the odds in favor of the event are 3:4.

2. When a single card is drawn from a deck of 52 cards, the odds against getting an ace card are 12:1; find the probability of selecting an ace.

3. Find the probability that an event *E* will not occur if the odds are 7:3 in favor of *E*.

4. When two dice are rolled, the odds in favor of getting a sum of 9 are 1:8; find the probability of getting a sum of 9.

5. Find the probability that an event *E* will not occur if the odds against the event *E* are 6:5.

**ANSWERS**

1. Let $A = 3$ and $B = 4$; then $P(E) = \frac{3}{3+4} = \frac{3}{7}$

2. Let $B = 12$ and $A = 1$; then $P(E) = \frac{1}{12+1} = \frac{1}{13}$

3. Let $A = 7$ and $B = 3$; then $P(E) = \frac{3}{7+3} = \frac{3}{10}$

4. Let $A = 1$ and $B = 8$; then $P(E) = \frac{1}{1+8} = \frac{1}{9}$

5. Let $B = 6$ and $A = 5$; then $P(\bar{E}) = \frac{6}{6+5} = \frac{6}{11}$

Expectation

When a person plays a slot machine, sometimes the person wins and other times—most often—the person loses. The question is, "How much will the person win or lose in the long run?" In other words, what is the person's expected gain or loss? Although an individual's exact gain or exact loss cannot be computed, the overall gain or loss of all people playing the slot machine can be computed using the concept of mathematical expectation.

Expectation or **expected value** is a **long-run average**. The expected value is also called the **mean**, and it is used in games of chance, insurance, and in other areas such as decision theory. The outcomes must be numerical in nature. The expected value of the outcome of a probability experiment can be found by multiplying each outcome by its corresponding probability and adding the results.

Formally defined, the expected value for the outcomes of a probability experiment is $E(X) = X_1 \cdot P(X_1) + X_2 \cdot P(X_2) + \cdots + X_n \cdot P(X_n)$, where the Xs correspond to the outcomes and the $P(X)$s are the corresponding probabilities of the outcomes.

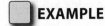

EXAMPLE

Find the expected value of the number of spots when a die is rolled.

SOLUTION

There are six outcomes when a die is rolled. They are 1, 2, 3, 4, 5, and 6, and each outcome has a probability of $\frac{1}{6}$ of occurring, so the expected value of the numbered spots is

$$E(X) = 1 \cdot \frac{1}{6} + 2 \cdot \frac{1}{6} + 3 \cdot \frac{1}{6} + 4 \cdot \frac{1}{6} + 5 \cdot \frac{1}{6} + 6 \cdot \frac{1}{6} = \frac{21}{6} = 3\frac{1}{2} \text{ or } 3.5.$$

The expected value is 3.5.

Now what does this mean? When a die is rolled, it is not possible to get 3.5 spots, but if a die is rolled say 100 times and the average of the spots is computed, that average should be close to 3.5 if the die is fair. In other words, 3.5 is the theoretical or long-run average. For example, if you rolled a die and were given $1 for each spot on each roll, sometimes you would win $1, $2, $3, $4, $5, or $6; however, on average, you would win $3.50 on each roll. So if you rolled the die 100 times, you would win on average $3.50 × 100 = $350. Now if you had to pay to play this game, you should pay $3.50 for each roll. That would make the game fair. If you paid more to play the game, say $4.00 each time you rolled the die, you would lose on average $0.50 on each roll. If you paid $3.00 to play the game, you would win an average $0.50 per roll.

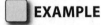

EXAMPLE

When three coins are tossed, find the expected value for the number of heads obtained.

SOLUTION

Consider the sample space when three coins are tossed.

HHH	HHT	HTH	THH	TTH	THT	HTT	TTT
|	\	|	/	\	|	/	|
3 heads		2 heads			1 head		0 heads

The probability of getting three heads is $\frac{1}{8}$. The probability of getting two heads is $\frac{3}{8}$. The probability of getting one head is $\frac{3}{8}$, and the probability of getting zero heads is $\frac{1}{8}$. The expected value for the number of heads is $E(X) = 3 \cdot \frac{1}{8} + 2 \cdot \frac{3}{8} + 1 \cdot \frac{3}{8} + 0 \cdot \frac{1}{8} = 1\frac{1}{2}$ or 1.5. Hence, the **average** number of heads obtained on each toss of three coins is 1.5.

In order to find the expected value for a gambling game, multiply the amount you win by the probability of winning that amount, and multiply the amount you lose by the probability of losing that amount, then add the results. Winning amounts are positive and losses are negative.

**EXAMPLE**

One thousand raffle tickets are sold for a prize of an entertainment center valued at $800. Find the expected value of the game if a person buys one ticket for $1.00.

**SOLUTION**

The problem can be set up as follows:

	Win	Lose
Gain, X	$799	−$1
Probability, P(X)	$\dfrac{1}{1000}$	$\dfrac{999}{1000}$

Since the person who buys a ticket does not get his or her $1 back, the net gain if he or she wins is $800 − $1 = $799. The probability of winning is one chance in 1000 since 1000 tickets are sold. The net loss is $1 denoted as negative and the chances of not winning are $\frac{1000-1}{1000}$ or $\frac{999}{1000}$. Now $E(X) = \$799 \cdot \frac{1}{1000} + (-\$1)\frac{999}{1000} = -0.2$ or −$0.20.

Here again it is necessary to realize that one cannot lose $0.20, but what this means is that the house (gambling establishment) makes $0.20 on every ticket sold. If a person purchased one ticket for raffles like this one over a long period of time, the person would lose **on average** $0.20 each time since he or she would win on average one time in 1000.

There is an alternative method that can be used to solve problems when tickets are sold or when people pay to play a game. In this case, multiply the overall gain by the probability of winning and subtract the cost of the ticket or the cost of playing the game. Using the information in the previous example, the solution looks like this:

$$E(X) = \$800 \cdot \frac{1}{1000} - \$1 = \$0.80 - \$1 = -\$0.20.$$

When the expected value is 0, the game is said to be fair. That is, there is a 50-50 chance of winning. When the expected value of a game is negative, the game is in favor of the house (i.e., the person or organization running the game). When the expected value of a game is positive, the game is in favor of the player. The last situation rarely ever happens unless the con man is not knowledgeable of probability theory.

 EXAMPLE

A player rolls a die. If he gets a 2, 3, or 6, he wins that amount of money. If he gets a 1, 4, or 5, he loses that amount of money. Find the expected value of the game.

 **SOLUTION**

$$E(X) = 2 \cdot \frac{1}{6} + 3 \cdot \frac{1}{6} + 6 \cdot \frac{1}{6} - 1 \cdot \frac{1}{6} - 4 \cdot \frac{1}{6} - 5 \cdot \frac{1}{6} = \frac{1}{6}.$$

Hence, the expected value of the game is $\frac{1}{6}$.

Expectation can be used to determine the average amount of money the house can make on each play of a gambling game. Consider the game called chuck-a-luck. A player pays \$1 and chooses a number from 1 to 6. Then three dice are tossed (usually in a cage). If the player's number comes up once, the player gets \$2. If it comes up twice, the player gets \$3, and if it comes up on all three dice, the player wins \$4. Con men like to say that the probability of any number coming up is $\frac{1}{6}$ on each die; therefore, each number has a probability of $\frac{3}{6}$ or $\frac{1}{2}$ of occurring, and if it occurs more than once, the player wins more money. Hence, the game is in favor of the player. This is not true. The next example shows how to compute the expected value for the game of chuck-a-luck.

EXAMPLE _____

Find the expected value for the game chuck-a-luck.

SOLUTION _____

There are $6 \times 6 \times 6 = 216$ outcomes in the sample space for three dice. The probability of winning on each die is $\frac{1}{6}$ and the probability of losing is $\frac{5}{6}$.

The probability that you win on all three dice is $\frac{1}{6} \cdot \frac{1}{6} \cdot \frac{1}{6} = \frac{1}{216}$.

The probability that you lose on all three dice is $\frac{5}{6} \cdot \frac{5}{6} \cdot \frac{5}{6} = \frac{125}{216}$.

The probability that you win on two dice is $\frac{1}{6} \cdot \frac{1}{6} \cdot \frac{5}{6} = \frac{5}{216}$, but this can occur in three different ways: win on the first die, win on the second die, and lose on the third die; or win on the first die, lose on the second die, and win on the third die; or lose on the first die, and win on the second and third die. Therefore, the probability of winning on two out of 3 dice is $3 \cdot \frac{5}{216} = \frac{15}{216}$.

The probability of winning on one die is $\frac{1}{6} \cdot \frac{5}{6} \cdot \frac{5}{6} = \frac{25}{216}$, and there are three different ways to win. Hence, the probability of winning on one die is $3 \cdot \frac{25}{216} = \frac{75}{216}$.

Now the expected value of the game is

X	\$3	\$2	\$1	−\$1
$P(X)$	$\dfrac{1}{216}$	$\dfrac{15}{216}$	$\dfrac{75}{216}$	$\dfrac{125}{216}$

$$E(X) = \$3 \cdot \frac{1}{216} + \$2 \cdot \frac{15}{216} + \$1 \cdot \frac{75}{216} - \$1 \cdot \frac{125}{216}$$

$$= -\frac{17}{216} \text{ or } -0.079 \text{ or about } -8 \text{ cents.}$$

Hence, on average, the house wins 8 cents on every game played by one player. If 5 people are playing, the house wins about $5 \times 8¢ = 40¢$ per game, on average.

PRACTICE

1. Five hundred tickets are sold at $2 each for a color television set worth $700. Find the expected value if a person purchased one ticket.

2. A lottery offers a $1000 prize. Find the expected value of the drawing if 1000 tickets are sold for $3.00 each and a person purchases one ticket.

3. A box contains three $1 bills, five $5 bills, and two $10 bills. If a person selects one bill at random, find the expected value of the draw.

4. If a person rolls a sum of seven in two tosses of a die, he wins $10. Find the expectation of the game if the person pays $4 to play.

5. If a person tosses two coins and gets two heads, the person wins $20. How much should the person pay if the game is to be fair?

 ANSWERS

1.

Gain, X	$698	−$2
Probability, $P(X)$	$\dfrac{1}{500}$	$\dfrac{499}{500}$

$$E(X) = \$698 \cdot \frac{1}{500} + (-\$2)\frac{499}{500} = -\$0.60$$

Alternate Solution

$$E(X) = \$700 \cdot \frac{1}{500} - \$2 = -\$0.60$$

2. $E(X) = \$997 \cdot \dfrac{1}{1000} - \$3.00 \cdot \dfrac{999}{1000} = -\$2.00$

3.

Gain, X	$10	$5	$1
Probability, $P(X)$	$\dfrac{2}{10}$	$\dfrac{5}{10}$	$\dfrac{3}{10}$

$$E(X) = \$10 \cdot \frac{2}{10} + \$5 \cdot \frac{5}{10} + \$1 \cdot \frac{3}{10} = \$4.80$$

4. There are 36 outcomes in the sample space.

Gain, X	$6	−$4
Probability, P(X)	$\dfrac{6}{36}$	$\dfrac{30}{36}$

$$E(X) = \$6 \cdot \frac{6}{36} + (-\$4) \cdot \frac{30}{36} = -\$2.33 \text{ (rounded)}$$

Alternate Solution

$$E(X) = \$10 \cdot \frac{6}{36} - \$4 = -\$2.33 \text{ (rounded)}$$

5. $P(\text{HH}) = \dfrac{1}{4}$

$$E(X) = 0 = \$20 \cdot \frac{1}{4} - x$$

$$x = \$5.00$$

A person should pay $5.00 to play the game.

PROBABILITY SIDELIGHT: Probability and Genetics

An Austrian botanist, Gregor Mendel (1822–1884), studied genetics and used probability theory to verify his results. Mendel lived in a monastery all of his adult life and based his research on the observation of plants. He published his results in an obscure journal and the results remained unknown until the beginning of the 20th century. At that time, his research was used by a mathematician, GH Hardy, to study human genetics.

Genetics is somewhat more complicated than what is presented here. However, what is important here is to explain how probability is used in genetics.

One of Mendel's studies was on the color of the seeds of pea plants. There were two colors, yellow and green. Mendel theorized that each egg cell and each pollen cell contained two color genes that split on fertilization. The offspring then contained one gene cell from each donor. There were three possibilities: pure yellow seeds, pure green seeds, and hybrid-yellow seeds. The pure yellow seeds contain two yellow genes. The pure green seeds contained two green genes. The hybrid-yellow seeds contain one yellow gene and one green gene. This seed was yellow since the yellow gene is dominant over the green gene. The green gene is said to be recessive.

Next consider the possibilities. If there are two pure yellow plants, then the results of fertilization will be YY as shown.

	Y	Y
Y	YY	YY
Y	YY	YY

Hence, $P(YY) = 1$.

The results of two pure green plants will be gg.

	g	g
g	gg	gg
g	gg	gg

Hence, $P(gg) = 1$.

What happens with a pure yellow plant and a pure green plant?

	g	g
Y	Yg	Yg
Y	Yg	Yg

Hence, $P(Yg) = 1$.

What happens with two hybrid yellow plants?

	Y	g
Y	YY	Yg
g	gY	gg

Hence, $P(YY) = \frac{1}{4}$, $P(Yg) = P(gY) = \frac{1}{2}$, and $P(gg) = \frac{1}{4}$.

What about a pure yellow plant and a hybrid yellow plant?

	Y	g
Y	YY	Yg
Y	YY	Yg

Hence, $P(YY) = \frac{1}{2}$ and $P(Yg) = \frac{1}{2}$.

What about a hybrid yellow plant and a pure green plant?

	g	g
Y	Yg	Yg
g	gg	gg

Hence, $P(Yg) = \frac{1}{2}$ and $P(gg) = \frac{1}{2}$.

This format can be used for other traits such as gender, eye color, etc. For example, for the gender of children, the female egg contains two X chromosomes, and the male are X and Y chromosomes. Hence, the results of fertilization are

	X	Y
X	XX	XY
X	XX	XY

Hence, $P(\text{female}) = P(XX) = \frac{1}{2}$ and $P(\text{male}) = P(XY) = \frac{1}{2}$.

Mendel could perform experiments and then compare the results with the theoretical probability of the outcomes in order to verify his hypotheses.

Summary

Odds are used to determine the payoffs in gambling games such as lotteries, horse races, and sports betting. Odds are computed from probabilities; however, probabilities can be computed from odds if the true odds are known.

Mathematical expectation can be thought of more or less as a long run average. If the game is played many times, the average of the outcomes or the payouts can be computed using mathematical expectation.

QUIZ

1. Three coins are tossed. What are the odds in favor of getting exactly 2 heads?
   A. 3:5
   B. 3:8
   C. 5:8
   D. 8:3

2. When two dice are rolled, what are the odds against getting a sum that is divisible by 3?
   A. 4:7
   B. 3:1
   C. 2:1
   D. 1:3

3. When a card is selected from a deck of 52 cards, what are the odds in favor of getting a red jack?
   A. 2:13
   B. 3:10
   C. 7:6
   D. 1:25

4. When a die is rolled, what are the odds in favor of getting a 2, 3, or 4?
   A. 2:3
   B. 1:1
   C. 3:2
   D. 2:1

5. On a roulette wheel, there are 38 numbers: 18 numbers are red and 18 numbers are black. Two are green. What are the odds in favor of getting a black or green number when the ball is rolled?
   A. 19:9
   B. 10:9
   C. 9:10
   D. 10:19

6. If the odds in favor of an event occurring are 4:9, what are the odds against the event occurring?
   A. 9:4
   B. 4:13
   C. 4:5
   D. 9:13

7. If the odds against an event occurring are 8:5, what are the odds in favor of the event occurring?

   A. 5:8
   B. 11:3
   C. 11:8
   D. 3:8

8. The probability of an event occurring is $\frac{3}{7}$. What are the odds in favor of the event occurring?

   A. 3:4
   B. 3:10
   C. 7:10
   D. 3:7

9. The probability of an event occurring is $\frac{8}{11}$. What are the odds against the event occurring?

   A. 8:11
   B. 8:3
   C. 3:8
   D. 3:19

10. What are the odds for a fair game?

    A. 0:0
    B. 1:1
    C. 2:1
    D. 1:2

11. When a game is fair, the expected value would be

    A. 1
    B. 0
    C. −1
    D. 0.5

12. When four coins are tossed, what is the expected value of the number of heads?

    A. 1
    B. 2
    C. 1.5
    D. 2.5

13. A special die is made with two 1s, three 2s, and one 3. What is the expected number of spots for one roll?

    A. $1\frac{5}{6}$
    B. $1\frac{2}{3}$
    C. 2
    D. $2\frac{1}{3}$

14. Four hundred raffle tickets are sold for $1.00 each. One prize of $50 is awarded. What is the expected value if a person purchases one ticket?

A. −$0.50

B. −$0.875

C. −$0.75

D. −$1.25

15. A box contains six $1 bills, two $5 bills, and two $10 bills. A person selects one bill at random and wins that bill. How much should the person pay to play the game if it is to be fair?

A. $4.00

B. $2.50

C. $3.60

D. $3.00

chapter 6

The Counting Rules

Since probability problems require knowing the total number of ways one or more events can occur, it is necessary to have a way to compute the number of outcomes in the sample spaces for a probability experiment. This is especially true when the number of outcomes is large. For example, when finding the probability of a specific poker hand, it is necessary to know the number of different possible ways five cards can be dealt from a 52-card deck. (This computation will be shown later in this chapter.)

In order to do the computation, we use the fundamental counting rule, the permutation rules, and the combination rule. The rules then can be used to compute the probability for events such as winning lotteries, getting a specific hand in poker, etc.

CHAPTER OBJECTIVES

In this chapter, you will learn

- How to determine the number of outcomes of an event using the fundamental counting rule, the permutation rules, and the combination rule
- How to find the probability of an event using these rules

The Fundamental Counting Rule

The first rule is called the **Fundamental Counting Rule**.

For a sequence of n events in which the first event can occur in k_1 ways and the second event can occur in k_2 ways and the third event can occur in k_3 ways, and so on, the total number of ways the sequence can occur is $k_1 \cdot k_2 \cdot k_3 \cdots k_n$.

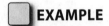**EXAMPLE**

At a picnic lunch, a patron can choose a hot dog or hamburger, French fries or baked potato, and soda, water, iced tea, or milk. How many different types of lunches can the patron choose if he selects one item from each category?

SOLUTION

The patron can choose one item from two types of sandwiches, one item from two types of potatoes, and one item from four kinds of drinks, so the patron has $2 \cdot 2 \cdot 4 = 16$ different ways to select a lunch.

**EXAMPLE**

A salesperson has to visit three cities. There are three ways to get from City A to City B and four ways to get from City B to City C. How many different ways can the sales person make the trip?

SOLUTION

$$3 \cdot 4 = 12$$

When determining the number of different ways a sequence of events can occur, it is necessary to know whether or not repetitions are permitted. The next two examples show the difference between the two situations.

**EXAMPLE**

A 5-digit identification card is to be issued to the employees of a medical facility.

a. How many different ID cards can be made if repetitions are permitted?

b. How many different ID cards can be made if no repetitions are permitted?

 **SOLUTION**

a. There are 10 digits (0 through 9), so if repetitions are permitted, then there could be $10 \cdot 10 \cdot 10 \cdot 10 \cdot 10 = 100{,}000$ different cards.

b. If repetitions are not permitted, there could be $10 \cdot 9 \cdot 8 \cdot 7 \cdot 6 = 30{,}240$ different cards.

PRACTICE

1. There are four types of blood types: A, B, AB, and O. Blood is either Rh$^+$ or Rh$^-$. How many different ways can a blood type be classified if the gender of the donor is included?

2. How many different types of identification cards consisting of six letters can be made from the first three letters of the alphabet if repetitions are allowed?

3. In order to paint a room, a person can select four different colors, two different types of paint, and three textures of paint. How many different combinations of paint are there?

4. A license plate consists of three letters and four digits. How many different plates can be made if repetitions are permitted? How many can be made if repetitions are not permitted?

5. How many different types of identification cards consisting of four letters can be made from the first five letters of the alphabet if repetitions are allowed?

ANSWERS

1. $4 \cdot 2 \cdot 2 = 16$ ways

2. $3 \cdot 3 \cdot 3 \cdot 3 \cdot 3 \cdot 3 = 729$ different cards

3. $4 \cdot 2 \cdot 3 = 24$ combinations

4. Repetitions permitted: $26 \cdot 26 \cdot 26 \cdot 10 \cdot 10 \cdot 10 \cdot 10 = 175{,}760{,}000$

   Repetitions not permitted: $26 \cdot 25 \cdot 24 \cdot 10 \cdot 9 \cdot 8 \cdot 7 = 78{,}624{,}000$

5. $5 \cdot 5 \cdot 5 \cdot 5 = 625$ different cards

Factorials

In mathematics there is a notation called **factorial notation**, which is used in the counting rule. This notation uses the exclamation point. Some examples of factorial notation are

$$5! = 5 \cdot 4 \cdot 3 \cdot 2 \cdot 1 = 120$$

$$2! = 2 \cdot 1 = 2$$

$$7! = 7 \cdot 6 \cdot 5 \cdot 4 \cdot 3 \cdot 2 \cdot 1 = 5040$$

$$1! = 1$$

Notice that factorial notation means to start with the number and find its product with all of the whole numbers less than the number (stopping at one). Formally defined, $n! = n \cdot (n-1) \cdot (n-2) \cdot \cdots \cdot 3 \cdot 2 \cdot 1$.

Factorial notation can be stopped at any time. For example,

$$5! = 5 \cdot 4! = 5 \cdot 4 \cdot 3!$$

$$9! = 9 \cdot 8! = 9 \cdot 8 \cdot 7!$$

In order to use the formulas in the rest of the chapter, it is necessary to know how to multiply and divide factorials. In order to multiply factorials, it is necessary to multiply them out and then multiply the products. For example,

$$3! \cdot 3! = 3 \cdot 2 \cdot 1 \cdot 3 \cdot 2 \cdot 1 = 36$$

Notice $3! \cdot 3! \neq 9!$ Since $9! = 362{,}880$.

 EXAMPLE

Find the product of $4! \cdot 6!$.

 **SOLUTION**

$$4! \cdot 6! = 4 \cdot 3 \cdot 2 \cdot 1 \cdot 6 \cdot 5 \cdot 4 \cdot 3 \cdot 2 \cdot 1 = 17{,}280.$$

Division of factorials is somewhat tricky. You can always multiply the factorials out and then divide the top number by the bottom number. For example,

$$\frac{8!}{6!} = \frac{8 \cdot 7 \cdot 6 \cdot 5 \cdot 4 \cdot 3 \cdot 2 \cdot 1}{6 \cdot 5 \cdot 4 \cdot 3 \cdot 2 \cdot 1} = \frac{40{,}320}{720} = 56$$

or

you can cancel out, as shown:

$$\frac{8!}{6!} = \frac{8 \cdot 7 \cdot \cancel{6!}}{\cancel{6!}} = 8 \cdot 7 = 56.$$

You cannot divide factorials directly:

$$\frac{8!}{4!} \neq 2! \text{ since } 8! = 40{,}320 \text{ and } 4! = 24 \text{ then } \frac{40{,}320}{24} = 1680.$$

EXAMPLE

Find the quotient $\dfrac{9!}{5!}$.

SOLUTION

$$\frac{9!}{5!} = \frac{9 \cdot 8 \cdot 7 \cdot 6 \cdot \cancel{5!}}{\cancel{5!}} = 3024.$$

Most scientific calculators have a factorial key. It is the key with "!". Also $0! = 1$ by definition.

PRACTICE

Find the value of each

1. $3!$

2. $8!$

3. $5!$

4. $1!$

5. $8! \cdot 3!$

6. $2! \cdot 4!$

7. $3! \cdot 6!$

8. $\dfrac{11!}{9!}$

9. $\dfrac{6!}{3!}$

10. $\dfrac{7!}{3!}$

ANSWERS

1. $3! = 3 \cdot 2 \cdot 1 = 6$

2. $8! = 8 \cdot 7 \cdot 6 \cdot 5 \cdot 4 \cdot 3 \cdot 2 \cdot 1 = 40{,}320$

3. $5! = 5 \cdot 4 \cdot 3 \cdot 2 \cdot 1 = 120$

4. $1! = 1$

5. $8! \cdot 3! = 8 \cdot 7 \cdot 6 \cdot 5 \cdot 4 \cdot 3 \cdot 2 \cdot 1 \cdot 3 \cdot 2 \cdot 1 = 241{,}920$

6. $2! \cdot 4! = 2 \cdot 1 \cdot 4 \cdot 3 \cdot 2 \cdot 1 = 48$

7. $3! \cdot 6! = 3 \cdot 2 \cdot 1 \cdot 6 \cdot 5 \cdot 4 \cdot 3 \cdot 2 \cdot 1 = 4320$

8. $\dfrac{11!}{9!} = \dfrac{11 \cdot 10 \cdot \cancel{9!}}{\cancel{9!}} = 110$

9. $\dfrac{6!}{3!} = \dfrac{6 \cdot 5 \cdot 4 \cdot \cancel{3!}}{\cancel{3!}} = 120$

10. $\dfrac{7!}{3!} = \dfrac{7 \cdot 6 \cdot 5 \cdot 4 \cdot \cancel{3!}}{\cancel{3!}} = 840$

Permutations

The second way to determine the number of outcomes of an event is to use the **permutation rules**. An arrangement of n distinct objects in a specific order is called a **permutation**. For example, if an art dealer had three paintings, say A, B, and C, to arrange in a row on a wall, there would be six distinct ways to display the paintings. They are

ABC	BAC	CAB
ACB	BCA	CBA

The total number of different ways can be found using the fundamental counting rule. There are three ways to select the first object, two ways to select the second object, and one way to select the third object. Hence, there are $3 \cdot 2 \cdot 1 = 6$ different ways to arrange three objects in a row on a shelf.

Another way to solve this kind of problem is to use permutations. The number of permutations of n objects using all the objects is $n!$.

 **EXAMPLE**

A person who is delivering flowers has seven customers to visit. How many different ways can this be done?

 SOLUTION

This is a permutation of seven people. Hence, $7! = 7 \cdot 6 \cdot 5 \cdot 4 \cdot 3 \cdot 2 \cdot 1 = 5040$ ways.

In the previous example, all the objects were used; however, in many situations only some of the objects are used. In this case, the **permutation rule** can be used.

The arrangement of n objects in a specific order using r objects at a time is called a **permutation** of n objects taking r objects at a time. It is written as $_nP_r$ and the formula is

$$_nP_r = \frac{n!}{(n-r)!}.$$

 EXAMPLE

How many different ways can an inspector visit three restaurants in a city that has a total of 10 restaurants?

SOLUTION

Since three restaurants are being selected from 10 restaurants and in a specific order, $n = 10$, $r = 3$. Hence, there are

$$_{10}P_3 = \frac{10!}{(10-3)!} = \frac{10!}{7!} = \frac{10 \cdot 9 \cdot 8 \cdot 7!}{7!} = 720 \text{ ways.}$$

 EXAMPLE

How many different ways can a researcher select four volunteers from 12 volunteers if each person is assigned to perform a different task?

SOLUTION

In this case, $n = 12$ and $r = 4$, so

$$_nP_r = {}_{12}P_4 = \frac{12!}{(12-4)!} = \frac{12!}{8!} = \frac{12 \cdot 11 \cdot 10 \cdot 9 \cdot 8!}{8!} = 11,880 \text{ ways.}$$

 **EXAMPLE**

A store owner has 15 items to advertise, and she can select only one different item to advertise each week for a 6-week period. How many different ways can this be done?

 SOLUTION

Here $n = 15$ and $r = 6$, so

$$_{15}P_6 = \frac{15!}{(15-6)!} = \frac{15!}{9!} = \frac{15 \cdot 14 \cdot 13 \cdot 12 \cdot 11 \cdot 10 \cdot \cancel{9!}}{\cancel{9!}} = 3{,}603{,}600.$$

In the preceding examples, all the objects were different, but when some of the objects are identical, the second permutation rule can be used.

The number of permutations of n objects when r_1 objects are identical, r_2 objects are identical, etc. is

$$\frac{n!}{r_1! r_2! \cdots r_p!}$$

where $r_1 + r_2 + \cdots + r_p = n$.

 **EXAMPLE**

How many different permutations can be made from the letters of the word **Alabama?**

 SOLUTION

There are four As, one L, one B, and one M; hence, $n = 7$, $r_1 = 4$, $r_2 = 1$, $r_3 = 1$, and $r_4 = 1$.

$$\frac{7!}{4! \cdot 1! \cdot 1! \cdot 1!} = \frac{7 \cdot 6 \cdot 5 \cdot \cancel{4!}}{\cancel{4!} \cdot 1! \cdot 1! \cdot 1!} = 210.$$

 **EXAMPLE**

An automobile dealer has two Fords, four Buicks, and five Dodges to place in the front row of his car lot. How many different ways by make of car can he display the automobiles?

SOLUTION

Let $n = 2 + 4 + 5 = 11$ automobiles; r_1 = 2 Fords, r_2 = 4 Buicks, and r_3 = 5 Dodges; then there are $\frac{11!}{2! \cdot 4! \cdot 5!} = \frac{11 \cdot 10 \cdot 9 \cdot 8 \cdot 7 \cdot 6 \cdot 5!}{2 \cdot 1 \cdot 4 \cdot 3 \cdot 2 \cdot 1 \cdot 5!} = 6930$ ways to display the automobiles.

PRACTICE

1. How many ways can eight books be arranged on a shelf?

2. How many different permutations can be made from the letters in the word Mississippi?

3. In how many different ways can a nurse select four patients from nine patients to visit in the next hour? The order of the visitations is important.

4. How many different signals using eight flags can be made if four are red, two are blue, and two are white?

5. How many different ways can a president, vice-president, secretary, and treasurer be selected from a club with 14 members?

ANSWERS

1. $8! = 8 \cdot 7 \cdot 6 \cdot 5 \cdot 4 \cdot 3 \cdot 2 \cdot 1 = 40{,}320$

2. $\dfrac{11!}{4! \cdot 4! \cdot 2! \cdot 1!} = \dfrac{11 \cdot 10 \cdot 9 \cdot 8 \cdot 7 \cdot 6 \cdot 5 \cdot 4!}{4! \cdot 4 \cdot 3 \cdot 2 \cdot 1 \cdot 2 \cdot 1 \cdot 1} = 34{,}650$

3. $_9P_4 = \dfrac{9!}{(9-4)!} = \dfrac{9!}{5!} = \dfrac{9 \cdot 8 \cdot 7 \cdot 6 \cdot 5!}{5!} = 3024$

4. $\dfrac{8!}{4! \cdot 2! \cdot 2!} = \dfrac{8 \cdot 7 \cdot 6 \cdot 5 \cdot 4!}{4! \cdot 2 \cdot 1 \cdot 2 \cdot 1} = 420$

5. $_{14}P_4 = \dfrac{14!}{(14-4)!} = \dfrac{14!}{10!} = \dfrac{14 \cdot 13 \cdot 12 \cdot 11 \cdot 10!}{10!} = 24{,}024$

Combinations

Sometimes when selecting objects, the order in which the objects are selected is not important. For example, when five cards are dealt in a poker game, the order in which you receive the cards is not important. When five balls are selected in a lottery, the order in which they are selected is not important. These situations differ from permutations (in which order is important) and are called combinations. A **combination** is a selection of objects without regard to the order in which they are selected.

Suppose two letters are selected from the four letters, A, B, C, and D. The different permutations are shown on the left and the different combinations are shown on the right.

Permutations				Combinations	
AB	BA	CA	DA	AB	BC
AC	BC	CB	DB	AC	BD
AD	BD	CD	DC	AD	CD

Notice that in a permutation AB differs from BA, but in a combination AB is the same as BA. The **combination rule** is used to find the number of ways to select objects without regard to order.

The number of r objects that can be selected from n objects without regard to order is

$$_nC_r = \frac{n!}{(n-r)!\,r!}$$

NOTE   *The symbol $_nC_r$ is used for combinations; however, some books use other symbols. Two of the most commonly used symbols are C_r^n or $\binom{n}{r}$.*

 **EXAMPLE**

How many ways can three objects be selected from seven objects without regard to order?

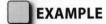 **SOLUTION**

Let $n = 7$ and $r = 3$, then

$$_7C_3 = \frac{7!}{(7-3)!\,3!} = \frac{7!}{4!\,3!} = \frac{7\cdot6\cdot5\cdot\cancel{4!}}{\cancel{4!}\cdot3\cdot2\cdot1} = 35 \text{ ways.}$$

 **EXAMPLE**

How many ways can five cards be selected from 52 cards? Disregard order.

SOLUTION

Let $n = 52$ and $r = 5$, then

$$_{52}C_5 = \frac{52!}{(52-5)!5!} = \frac{52!}{47!5!} = \frac{52 \cdot 51 \cdot 50 \cdot 49 \cdot 48 \cdot \cancel{47!}}{\cancel{47!} \cdot 5 \cdot 4 \cdot 3 \cdot 2 \cdot 1} = 2{,}598{,}960 \text{ ways.}$$

EXAMPLE

How many ways can a person download nine songs to his iPod if he selects three rock songs from five rock songs, two rap songs from six rap songs, and four old-ies from seven oldies? Order is not important.

SOLUTION

In this case, the person must select three rock songs from five rock songs, two rap songs from six rap songs, and four oldies from seven oldies. Since the word "and" is used, you must multiply.

$$_5C_3 \cdot _6C_2 \cdot _7C_4 = \frac{5!}{(5-3)!3!} \cdot \frac{6!}{(6-2)!2!} \cdot \frac{7!}{(7-4)!4!}$$

$$= \frac{5!}{2!3!} \cdot \frac{6!}{4!2!} \cdot \frac{7!}{3!4!} = \frac{5 \cdot 4 \cdot \cancel{3!}}{\cancel{3!} \cdot 2 \cdot 1} \cdot \frac{6 \cdot 5 \cdot \cancel{4!}}{\cancel{4!} \cdot 2 \cdot 1} \cdot \frac{7 \cdot 6 \cdot 5 \cdot \cancel{4!}}{3 \cdot 2 \cdot 1 \cdot \cancel{4!}}$$

$$= 10 \cdot 15 \cdot 35 = 5250 \text{ ways.}$$

PRACTICE

1. A census worker has to visit 11 houses. How many ways can she select four houses to visit before lunch? Disregard order.

2. How many ways can a jury of 12 people be seated from 15 people?

3. How many ways can a committee of five people be selected from eight people?

4. How many ways can two men and two women be selected from eight men and six women?

5. How many ways can three hockey players and four soccer players be se-lected from 7 hockey players and 10 soccer players?

 ANSWERS

1. $\displaystyle {}_{11}C_4 = \frac{11!}{(11-4)!4!} = \frac{11!}{7!4!} = \frac{11 \cdot 10 \cdot 9 \cdot 8 \cdot 7!}{7! \cdot 4 \cdot 3 \cdot 2 \cdot 1} = 330$

2. $\displaystyle {}_{15}C_{12} = \frac{15!}{(15-12)!12!} = \frac{15!}{3!12!} = \frac{15 \cdot 14 \cdot 13 \cdot 12!}{3 \cdot 2 \cdot 1 \cdot 12!} = 455$

3. $\displaystyle {}_{8}C_5 = \frac{8!}{(8-5)!5!} = \frac{8!}{3!5!} = \frac{8 \cdot 7 \cdot 6 \cdot 5!}{3 \cdot 2 \cdot 1 \cdot 5!} = 56$

4. $\displaystyle {}_{8}C_2 \cdot {}_{6}C_2 = \frac{8!}{(8-2)!2!} \cdot \frac{6!}{(6-2)!2!} = \frac{8!}{6!2!} \cdot \frac{6!}{4!2!}$

    $\displaystyle = \frac{8 \cdot 7 \cdot 6!}{6! \cdot 2 \cdot 1} \cdot \frac{6 \cdot 5 \cdot 4!}{4! \cdot 2 \cdot 1} = 28 \cdot 15 = 420$

5. $\displaystyle {}_{7}C_3 \cdot {}_{10}C_4 = \frac{7!}{(7-3)!3!} \cdot \frac{10!}{(10-4)!4!} = \frac{7!}{4!3!} \cdot \frac{10!}{4!6!}$

    $\displaystyle = \frac{7 \cdot 6 \cdot 5 \cdot 4!}{4! \cdot 3 \cdot 2 \cdot 1} \cdot \frac{10 \cdot 9 \cdot 8 \cdot 7 \cdot 6!}{4 \cdot 3 \cdot 2 \cdot 1 \cdot 6!} = 35 \cdot 210 = 7350$

Probability and the Counting Rules

A wide variety of probability problems can be solved using the counting rules.

EXAMPLE

Find the probability that a committee of four women and two men are randomly selected from six women and five men.

 SOLUTION

There are ${}_{6}C_4 \cdot {}_{5}C_2$ ways of selecting the women and the men and ${}_{11}C_6$ total ways of selecting 6 people from 11 people. Hence,

$$ {}_{6}C_4 \cdot {}_{5}C_2 = \frac{6!}{(6-4)!4!} \cdot \frac{5!}{(5-2)!2!} = \frac{6!}{2!4!} \cdot \frac{5!}{3!2!} $$

$$ = \frac{6 \cdot 5 \cdot 4!}{2 \cdot 1 \cdot 4!} \cdot \frac{5 \cdot 4 \cdot 3!}{3! \cdot 2 \cdot 1} = 15 \cdot 10 = 150 $$

$$ {}_{11}C_6 = \frac{11!}{(11-6)!6!} = \frac{11!}{5!6!} = \frac{11 \cdot 10 \cdot 9 \cdot 8 \cdot 7 \cdot 6!}{5 \cdot 4 \cdot 3 \cdot 2 \cdot 1 \cdot 6!} = 462 $$

So the probability is $\frac{150}{462} = \frac{25}{77}$.

EXAMPLE

Find the probability of getting a full house when five cards are dealt from a 52-card deck.

SOLUTION

A full house means three cards of one denomination and two cards of a different denomination. So there are $_{13}C_2$ ways to choose the two denominations. Also, the full house of three kings and two queens is different from the full house of three queens and two kings, so there are two ways to get the suits. Finally, there are $_4C_3$ ways to get three cards and $_4C_2$ ways to get the two cards. Hence, the number of different ways to get a full house is $2 \cdot {}_{13}C_2 \cdot {}_4C_3 \cdot {}_4C_2 = 2 \cdot \frac{13!}{11!2!} \cdot \frac{4!}{1!3!} \cdot \frac{4!}{2!2!} = 3744$. There are $_{52}C_5$ ways to be dealt 5 cards. So $_{52}C_5 = 2{,}598{,}960$. The probability of being dealt a full house is $\frac{3{,}744}{2{,}598{,}960} \approx 0.0014$.

EXAMPLE

The red cards numbered 2 through 10 and the black cards numbered 2 through 10 are placed in a bag. If four cards are selected at random, find the probability that two are red and two are black.

SOLUTION

There are $_9C_2$ ways to select two red cards and $_9C_2$ ways to select two black cards, so the number of ways to select two red cards and two black cards is

$$_9C_2 \cdot {}_9C_2 = \frac{9!}{(9-2)!2!} \cdot \frac{9!}{(9-2)!2!} = \frac{9!}{7!2!} \cdot \frac{9!}{7!2!} = \frac{9 \cdot 8 \cdot 7!}{7!2!} \cdot \frac{9 \cdot 8 \cdot 7!}{7!2!} = 1296.$$

The number of ways of selecting four cards from 18 cards is

$$_{18}C_4 = \frac{18!}{(18-4)!} = \frac{18!}{14!4!} = \frac{18 \cdot 17 \cdot 16 \cdot 15 \cdot 14!}{14! \cdot 4 \cdot 3 \cdot 2 \cdot 1} = 3060.$$

Hence, the probability of selecting two red cards and two black cards is

$$P(2 \text{ red and } 2 \text{ black}) = \frac{1296}{3060} = \frac{36}{85} \approx 0.42.$$

PRACTICE

1. A box contains 10 resistors, four of which are defective. If three resistors are selected at random, find the probability that all three are defective.

2. Find the probability of selecting two science books and three math books from six science books and five math books.

3. At a political rally, there are 10 Democrats, eight Republicans, and three independents. If three people are selected at random, find the probability that one of each party is selected.

4. If three dice are rolled, find the probability of getting a sum of 6.

5. Five students line up at random for a picture. Find the probability that they line up according to height. Assume no two students are of the same height.

ANSWERS

1. There are $_{10}C_3$ ways of selecting three resistors from 10 resistors.

$$_{10}C_3 = \frac{10!}{(10-3)!3!} = \frac{10!}{7!3!} = \frac{10 \cdot 9 \cdot 8 \cdot 7!}{7! \cdot 3 \cdot 2 \cdot 1} = 120.$$

There are $_4C_3$ ways to select three defective resistors from four defective resistors.

$$_4C_3 = \frac{4!}{(4-3)!3!} = \frac{4!}{1! \cdot 3!} = \frac{4 \cdot 3!}{1 \cdot 3!} = 4$$

$P(3 \text{ defective resistors}) = \frac{4}{120} = \frac{1}{30}$.

2. There are $_{11}C_5$ ways of selecting five books from 11 books.

$$_{11}C_5 = \frac{11!}{(11-5)!5!} = \frac{11!}{6!5!} = \frac{11 \cdot 10 \cdot 9 \cdot 8 \cdot 7 \cdot 6!}{6! \cdot 5 \cdot 4 \cdot 3 \cdot 2 \cdot 1} = 462.$$

There are $_6C_2$ ways to select two science books from six science books and $_5C_3$ ways to select three math books from five math books. So there are $_6C_2 \cdot {_5C_3}$ ways to select these books.

$$_6C_2 = \frac{6!}{(6-2)!2!} = \frac{6!}{4!2!} = \frac{6 \cdot 5 \cdot 4!}{4! \cdot 2 \cdot 1} = 15$$

$$_5C_3 = \frac{5!}{(5-3)!3!} = \frac{5!}{2!3!} = \frac{5 \cdot 4 \cdot 3!}{3! \cdot 2 \cdot 1} = 10$$

$$15 \cdot 10 = 150$$

$P(2 \text{ science books and 3 math books}) = \frac{150}{462} = \frac{25}{77}$.

3. There are $_{21}C_3$ ways to select three people from 21 people.

$$_{21}C_3 = \frac{21!}{(21-3)!3!} = \frac{21!}{18!3!} = \frac{21\cdot 20\cdot 19\cdot \cancel{18!}}{\cancel{18!}\cdot 3\cdot 2\cdot 1} = 1330.$$

There are $_{10}C_1$ ways to select the Democrat, $_8C_1$ ways to select the Republican, and $_3C_1$ ways to select the Independent.

$$_{10}C_1 = \frac{10!}{(10-1)!1!} = \frac{10!}{9!1!} = \frac{10\cdot \cancel{9!}}{\cancel{9!}\cdot 1} = 10$$

$$_8C_1 = \frac{8!}{(8-1)!1!} = \frac{8!}{7!1!} = \frac{8\cdot \cancel{7!}}{\cancel{7!}} = 8$$

$$_3C_1 = \frac{3!}{(3-1)!1!} = \frac{3!}{2!1!} = \frac{3\cdot \cancel{2!}}{\cancel{2!}\cdot 1} = 3$$

So there are $10\cdot 8\cdot 3 = 240$ ways to select one person from each party.

$P(\text{selecting 1 person from each party}) = \frac{240}{1330} = \frac{24}{133}$.

4. There are $6\cdot 6\cdot 6 = 216$ members of the sample space since each die can show six different numbers. There are 10 ways to get a sum of 6. They are (4,1,1), (1,4,1), (1,1,4), (3,2,1), (3,1,2), (1,3,2), (2,3,1), (2,1,3), (1,2,3), and (2,2,2).

$P(\text{sum of 6}) = \frac{10}{216} = \frac{5}{108}$.

5. There are 5! ways for the students to line up. 5! = 120. Now there are two ways to line up according to height, short to tall or tall to short.

$P(\text{line up according to height}) = \frac{2}{120} = \frac{1}{60}$.

PROBABILITY SIDELIGHT: The Classical Birthday Problem

What do you think the chances are that in a classroom of 23 students, two students would have the same birthday (day and month)? Most people would think the probability is very low since there are 365 days in a year; however, the probability is slightly greater than 50%! Furthermore, as the number of students increases, the probability increases very rapidly. For example, if there are 30 students in the room, there is a 70% chance that two students will have the same birthday, and when there are 50 students in the room, the probability jumps to 97%!

The classical birthday problem uses permutations and the probability rules. It must be assumed that all birthdays are equally likely. This is not necessarily true, but it has little effect on the solution. The way to solve the problem is to find the probability that no two people have the same birthday and subtract it from one. Recall $P(E)=1-P(\bar{E})$.

For example, suppose that there were only three people in the room. Then the probability that each would have a different birthday would be

$$\left(\frac{365}{365}\right)\cdot\left(\frac{364}{365}\right)\cdot\left(\frac{363}{365}\right)=\frac{_{365}P_3}{(365)^3}=0.992.$$

The reasoning here is that the first person could be born on any day of the year. Now if the second person would have a different birthday, there are 364 days left, so the probability that the second person was born on a different day is $\frac{364}{365}$. The reasoning is the same for the next person. Now since the probability is 0.992 that the three people have different birthdays, the probability that any two have the same birthday is $1 - 0.992 = 0.008$ or 0.8%.

In general, in a room with k people, the probability that at least two people will have the same birthday is $1-\frac{_{365}P_k}{365^k}$.

In a room with 23 students, the probability that at least two students will have the same birthday is $1-\frac{_{365}P_{23}}{365^{23}}=0.507$ or 50.7%.

It is interesting to note that two presidents, James K. Polk and Warren G. Harding, were both born on November 2. Also, John Adams and Thomas Jefferson both died on July 4. What is even more unusual is that they both died on the same day of the same year, July 4, 1826. Another president, James Monroe, also died on July 4, but the year was 1831.

Summary

In order to determine the number of outcomes of events, the fundamental counting rule, the permutation rules, and the combination rule can be used. The difference between a permutation and a combination is that for a permutation, the order or arrangement of the objects is important. For example, order is important in phone numbers, identification tags, social security numbers, license plates, etc. Order is not important when selecting objects from a group. Many probability problems can be solved by using the counting rules to determine the number of ways the events can occur.

QUIZ

1. **The value of 8! is**
   A. 80,235
   B. 4096
   C. 40,320
   D. 6720

2. **The value of 0! is**
   A. 0
   B. 1
   C. 10
   D. 100

3. **The value of $_6P_4$ is**
   A. 720
   B. 30
   C. 120
   D. 360

4. **The value of $_7C_3$ is**
   A. 35
   B. 21
   C. 360
   D. 120

5. **The number of 4-digit identification cards that can be made if repetitions are not allowed is**
   A. 256
   B. 720
   C. 5040
   D. 24

6. **How many different ways can a person select one book from six novels, one book from three biographies, and one book from five self-help books?**
   A. 14
   B. 3
   C. 30
   D. 90

7. **How many ways can five different printers be displayed in a row on a shelf?**
   A. 120
   B. 5040
   C. 5
   D. 3125

8. If a board of directors consists of 18 people, how many ways can a chief executive officer, a director, a treasurer, and a secretary be selected?

   A. 40
   B. 5040
   C. 3060
   D. 73,440

9. How many different flag signals, using all the flags at once, can be made from six black flags, four white flags, and three blue flags?

   A. 13
   B. 4620
   C. 60,060
   D. 27,720

10. How many ways can five iPods be selected for testing from nine iPods? (Order is not important.)

    A. 45
    B. 120
    C. 126
    D. 240

11. How many ways can a swim team of four men and four women be selected from nine men and nine women?

    A. 16
    B. 126
    C. 252
    D. 15,876

12. A phone extension consists of three digits. If all possible digits each have an equal probability of being selected with repetition, what is the probability that the extension consists of the digits 1, 2, and 3 in any order if repetitions here are not allowed?

    A. 0.050
    B. 0.006
    C. 0.233
    D. 0.125

13. Three cards are selected at random; what is the probability that they will all be face cards? (Round to three decimal places.)

    A. 0.002
    B. 0.034
    C. 0.010
    D. 0.127

14. At a used book sale, there are six novels and four biographies. If a person selects four books at random, what is the probability that the person selects three novels and one biography? (Round to three decimal places.)

    A. 0.137
    B. 0.114
    C. 0.437
    D. 0.381

15. To win a lottery, a person must select three numbers in any order from 10 numbers. Repetitions are not allowed. What is the probability that the person wins? (Round the answer to four decimal places.)

    A. 0.0055
    B. 0.0033
    C. 0.0062
    D. 0.0083

chapter 7

The Binomial Distribution

Many probability problems involve assigning probabilities to the outcomes of a probability experiment. These probabilities and the corresponding outcomes make up a *probability distribution*. There are many different probability distributions. One special probability distribution is called the *binomial distribution*. The binomial distribution has many uses such as in gambling, in inspecting parts, and in other areas.

CHAPTER OBJECTIVES

In this chapter, you will learn

- The characteristics of a binomial distribution
- How to picture a binomial distribution
- How to find the probability of an outcome of a binomial distribution
- How to find the mean, variance, and standard deviation for the variable of a binomial distribution

Discrete Probability Distributions

In mathematics, a **variable** can assume different values. For example, if one records the temperature outside every hour for a 24-hour period, temperature is considered a variable since it assumes different values. Variables whose values are due to chance are called **random variables**. When a die is rolled, the value of the spots on the face up occurs by chance; hence, the number of spots on the face up on the die is considered to be a random variable. The outcomes of a die are 1, 2, 3, 4, 5, and 6, and the probability of each outcome occurring is $\frac{1}{6}$. The outcomes and their corresponding probabilities can be written in a table, as shown, and make up what is called a probability distribution.

Value, x	1	2	3	4	5	6
Probability, $P(x)$	$\frac{1}{6}$	$\frac{1}{6}$	$\frac{1}{6}$	$\frac{1}{6}$	$\frac{1}{6}$	$\frac{1}{6}$

A **probability distribution** consists of the values of a random variable and their corresponding probabilities. There are two kinds of probability distributions. They are *discrete probability distributions* and *continuous probability distributions*.

A discrete probability distribution uses discrete variables. A **discrete** variable has a countable number of values (countable means values of 0, 1, 2, 3, etc.). For example, when four coins are tossed, the outcomes for the number of heads obtained are 0, 1, 2, 3, and 4. When a single die is rolled, the outcomes are 1, 2, 3, 4, 5, and 6. These are examples of discrete variables.

A continuous probability distribution consists of continuous variables. A **continuous** variable has an infinite number of values between any two values. For example, temperature is a continuous variable since the variable of temperature can assume any value between 10° and 20° or any other two temperatures for that matter. Height and weight are continuous variables. Of course, we are limited by our measuring devices and values of continuous variables are usually rounded. Continuous variables are measured.

**EXAMPLE**

Construct a discrete probability distribution for the number of heads when three coins are tossed.

SOLUTION

Recall that the sample space for tossing three coins is

TTT, TTH, THT, HTT, HHT, HTH, THH, and HHH.

The outcomes can be arranged according to the number of heads, as shown.

0 heads	TTT		
1 head	TTH,	THT,	HTT
2 heads	THH,	HTH,	HHT
3 heads	HHH		

Finally, the outcomes and corresponding probabilities can be written in a table, as shown.

Outcome, x	0	1	2	3
Probability, $P(x)$	$\frac{1}{8}$	$\frac{3}{8}$	$\frac{3}{8}$	$\frac{1}{8}$

The sum of the probabilities of a probability distribution will always be equal to 1.

A discrete probability distribution can also be shown graphically by labeling the x-axis with the values of the outcomes and letting the values on the y-axis represent the probabilities for the outcomes. The graph for the discrete probability distribution of the number of heads occurring when three coins are tossed is shown in Figure 7-1.

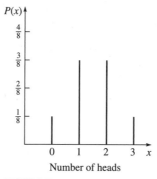

FIGURE 7-1

There are many kinds of discrete probability distributions; however, the distribution of the number of heads when three coins are tossed is a special kind of distribution called a *binomial distribution*.

A binomial distribution is obtained from a probability experiment called a **binomial experiment**. The experiment must satisfy these conditions:

1. Each trial can have only two outcomes or outcomes that can be reduced to two outcomes. The outcomes are usually considered as a success or a failure.

2. There is a fixed number of trials.

3. The outcomes of each trial are independent of each other.

4. The probability of a success must remain the same for each trial.

EXAMPLE

Explain why the probability experiment of tossing three coins is a binomial experiment.

 **SOLUTION**

In order to be a binomial experiment, the probability experiment must satisfy the four conditions explained previously.

1. There are only two outcomes for each trial, head and tail. Depending on the situation, either heads or tails can be defined as a success and the other as a failure.

2. There is a fixed number of trials. In this case, there are three trials since three coins are tossed or one coin is tossed three times.

3. The outcomes are independent since tossing one coin does not affect the outcome of the other two tosses.

4. The probability of a success (say heads) is $\frac{1}{2}$ and it does not change.

Hence, the experiment meets the conditions of a binomial experiment.

Now consider rolling a die. Since there are six outcomes, it cannot be considered a binomial experiment. However, it can be made into a binomial experiment by considering the outcome of getting five spots a success and every other outcome as a failure.

In order to determine the probability of exactly x successes in n trials of a probability experiment, the following formula can be used.

$$_nC_x \cdot (p)^x \cdot (1-p)^{n-x}$$

where n = the total number of trials
   x = the number of successes $(1, 2, 3, \ldots, n)$
   p = the probability of a success

The formula has three parts: $_nC_x$ determines the number of ways a success can occur. $(p)^x$ is the probability of getting x successes, and $(1-p)^{n-x}$ is the probability of getting $n - x$ failures.

 EXAMPLE

A coin is tossed five times. Find the probability of getting three heads and two tails in any given order.

 SOLUTION

Since the coin is tossed five times, $n = 5$. The probability of getting a head (success) is $\frac{1}{2}$, so $p = \frac{1}{2}$ and the probability of getting a tail (failure) is $1 - \frac{1}{2} = \frac{1}{2}$; $x = 3$ since the problem asks for three heads.

$$(n - x) = 5 - 3 = 2.$$

Hence,

$$P(3 \text{ heads}) = {_5C_3} \cdot \left(\frac{1}{2}\right)^3 \left(\frac{1}{2}\right)^2$$

$$= \frac{5 \cdot 4 \cdot \cancel{3!}}{\cancel{3!} \cdot 2 \cdot 1}\left(\frac{1}{8}\right)\left(\frac{1}{4}\right)$$

$$= \frac{10}{32} = \frac{5}{16}.$$

 EXAMPLE

A die is rolled four times; find the probability of getting exactly two 5s.

 SOLUTION

Since we are rolling a die 4 times, $n = 4$. The probability of getting a 5 is $\frac{1}{6}$. The probability of not getting a 5 is $1 - \frac{1}{6}$ or $\frac{5}{6}$. Since a success is getting a 5, and since two 5s are desired, we have $x = 2$ and $n - x = 4 - 2 = 2$. Hence,

$$P(\text{exactly two 5s}) = {}_4C_2 \cdot \left(\frac{1}{6}\right)^2 \cdot \left(\frac{5}{6}\right)^2$$

$$= \frac{4 \cdot 3 \cdot 2!}{2! \cdot 2 \cdot 1}\left(\frac{1}{36}\right)\left(\frac{25}{36}\right)$$

$$= \frac{\cancel{4}^1 \cdot \cancel{3}^1 \cdot 1 \cdot 25}{2 \cdot \cancel{36}^{9^3} \cdot 36} = \frac{25}{216} \approx 0.116 = 11.6\%.$$

About 11.6% of the time, exactly two 5s will occur.

EXAMPLE

An archer hits the bull's eye 60% of the time. If he shoots five arrows, find the probability that he will hit the target at most three times.

SOLUTION

When the problem says at most three times, this means 0, 1, 2, or 3 times. This means that you need to find each probability, and then add the results. In this case, the probability of a success is 0.6, and the probability of a failure is $1 - 0.6 = 0.4$.

$$P(x = 0) = {}_5C_0(0.6)^0(0.4)^5 = \frac{5!}{(5-0)!0!} \cdot 1 \cdot (0.4)^5$$

$$= 1 \cdot 1 \cdot (0.01024) \approx 0.0102$$

$$P(x = 1) = {}_5C_1(0.6)^1(0.4)^4 = 5(0.6)(0.0256) = 0.0768$$

$$P(x = 2) = {}_5C_2(0.6)^2(0.4)^3 = (10)(0.36)(0.064) = 0.2304$$

$$P(x = 3) = {}_5C_3(0.6)^3(0.4)^2 = (10)(0.216)(0.16) = 0.3456$$

Thus, the desired probability is sum of the above which is approximately equal to 0.443.

NOTE *The probabilities have been rounded to three decimal places. Hence, the probability that he will hit the target at most three times is 0.0102 + 0.0768 + 0.2304 + 0.3456 = 0.663 or 66.3%.*

As stated previously, in order to construct a probability distribution, the following formula is used:

$${}_nC_x\, p^x(1 - p)^{n-x}, \text{ where } x = 1, 2, 3, \ldots, n.$$

The next example shows how to use the formula.

EXAMPLE

A box contains three red balls and seven white balls. A ball is selected, its color is noted, and then it is replaced. Construct a probability distribution for the number of red balls that will occur if the experiment is done three times.

SOLUTION

In this case, a ball is selected and replaced three times, so $n = 3$. The probability of getting a red ball is 0.3, and one can get $x = 0, 1, 2,$ or 3 red balls.

For $x = 0$, $_3C_0(0.3)^0(0.7)^3 = 0.3430$

For $x = 1$, $_3C_1(0.3)^1(0.7)^2 = 0.4410$

For $x = 2$, $_3C_2(0.3)^2(0.7)^1 = 0.1890$

For $x = 3$, $_3C_3(0.3)^3(0.7)^0 = 0.0270$

Hence, the probability distribution is

Number of red balls, x	0	1	2	3
Probability, $P(x)$	0.3430	0.4410	0.1890	0.0270

Still Struggling

Many statistics books have tables of computed probabilities for binomial variables. Also, several types of calculators can be used to find the specific probabilities for binomial variables.

PRACTICE

1. A recent survey found that 10% of students in a certain school district drop out before they graduate. If 20 students are selected at random, find the probability that 5 will drop out before they graduate.

2. On a 10-question multiple choice exam with five possible choices for each question, a student guesses the answer on each question. Find the probability that the student gets seven correct.

3. An 8-sided die is rolled five times. Find the probability of getting exactly 3 sixes.

4. Six coins are tossed; find the probability of getting at most two heads.

5. A deck of cards consists of 12 red cards and eight black cards. The cards are mixed up and two cards are selected. Find the probability that at least one of the cards is red.

ANSWERS

1. $n = 20, x = 5, p = 0.10$

$$P(\text{exactly 5 will drop out}) = {}_{20}C_5 (0.10)^5 (0.90)^{15} = 0.0319.$$

2. $n = 10, x = 7, p = 0.20$ (1 correct answer in 5 choices)

$$P(\text{exactly 7 correct}) = {}_{10}C_7 (0.20)^7 (0.8)^3 = 0.0008.$$

3. $n = 5, x = 3, p = \frac{1}{8} = 0.125$

$$P(\text{exactly 3 sixes}) = {}_5C_3 (0.125)^3 (0.875)^2 = 0.01495.$$

4. $n = 6, x = 0, 1, 2, p = \frac{1}{2}$

For 0 heads, ${}_6P_0 \left(\dfrac{1}{2}\right)^0 \left(\dfrac{1}{2}\right)^6 = 0.0156$

For 1 head, ${}_6P_1 \left(\dfrac{1}{2}\right)^1 \left(\dfrac{1}{2}\right)^5 = 0.0938$

For 2 heads, ${}_6P_2 \left(\dfrac{1}{2}\right)^2 \left(\dfrac{1}{2}\right)^4 = 0.2344$

$0.0156 + 0.0938 + 0.2344 = 0.3438.$

5. $n = 2, x = 1$ or $2, p = \frac{12}{20} = 0.6$

For $x = 1, {}_2C_1 (0.6)^1 (0.4)^1 = 0.4800$

For $x = 2, {}_2C_2 (0.6)^2 (0.4)^0 = 0.3600$

$0.4800 + 0.3600 = 0.84.$

The Mean and Standard Deviation for a Binomial Distribution

Suppose you roll a die many times and record the number of 3s you obtain. Is it possible to predict ahead of time the average number of 3s you will obtain? The answer is "Yes." It is called the **expected value** or the **mean** of a binomial distribution. This mean can be found by using the formula mean $(\mu) = np$, where n is the number of times the experiment is repeated and p is the probability of a success. The symbol for the mean is the Greek letter μ (mu).

 **EXAMPLE**

A die is tossed 180 times and the number of 3s obtained is recorded. Find the mean or expected number of 3s.

SOLUTION

$n = 180$ and $p = \frac{1}{6}$ since there is one chance in six to get a 3 on each roll.

$$\mu = n \cdot p = 180 \cdot \frac{1}{6}$$
$$= 30$$

Hence, one would expect on average 30 threes.

NOTE *The probability of getting exactly 30 threes is small, but if you repeated the experiment many times, the average would approach 30.*

 **EXAMPLE**

Eight cards are selected from a deck and each card is replaced before the next one is drawn. Find the average number of diamonds.

SOLUTION

In this case, $n = 8$ and $p = \frac{13}{52}$ or $\frac{1}{4}$ since there are 13 diamonds and a total of 52 cards. The mean is

$$\mu = n \cdot p$$
$$= 8 \cdot \frac{1}{4}$$
$$= 2.$$

Hence, on average, we would expect two diamonds in a set of eight cards.

Statisticians are not only interested in the average of the outcomes of a probability experiment but also in how the results of a probability experiment vary from trial to trial. Suppose we roll a die 180 times and record the number of threes obtained. We know that we would expect to get about 30 threes. Now what if the experiment was repeated again and again? In this case, the number of threes obtained each time would not always be 30 but would vary about the mean of 30. For example, we might get 28 threes one time and 34 threes the next time, etc. How can this variability be explained? Statisticians use a measure called the **standard deviation**. When the standard deviation of a variable is large, the individual values of the variable are spread out from the mean of the distribution. When the standard deviation of a variable is small, the individual values of the variable are close to the mean.

The formula for the standard deviation for a binomial distribution is standard deviation $\sigma = \sqrt{np(1-p)}$. The symbol for the standard deviation is the Greek letter σ (sigma).

 **EXAMPLE**

A die is rolled 180 times. Find the standard deviation of the number of threes.

SOLUTION

$$n = 180, p = \frac{1}{6}, 1 - p = 1 - \frac{1}{6} = \frac{5}{6}$$

$$\sigma = \sqrt{np(1-p)}$$

$$= \sqrt{180 \cdot \frac{1}{6} \cdot \frac{5}{6}}$$

$$= \sqrt{25}$$

$$= 5$$

The standard deviation is 5.

Now what does this tell us?

Roughly, most of the values fall within two standard deviations of the mean.

$$\mu - 2\sigma < \text{most values} < \mu + 2\sigma$$

In the die example, we can expect most values to fall between

$$30 - 2 \cdot 5 < \text{most values} < 30 + 2 \cdot 5$$

$$30 - 10 < \text{most values} < 30 + 10$$

$$20 < \text{most values} < 40$$

In this case, if we did the experiment many times, we would expect between 20 and 40 threes most of the time. This is an approximate "range of values."

Suppose we rolled a die 180 times and we got only 5 threes, what can be said? It can be said that this is an unusually small number of threes. It can happen by chance, but not very often. We might want to consider some other possibilities. Perhaps the die is loaded or perhaps the die has been manipulated by the person rolling it! But remember it *can* happen by chance.

EXAMPLE

It is estimated that 25% of people eat at a fast-food restaurant each day. If 200 people are randomly selected on a particular day, find the mean and standard deviation of the number of people who will eat at a fast-food restaurant. Find the approximate range of values.

SOLUTION

$n = 200, p = 0.25, q = 1 - 0.25 = 0.75$

Mean:
$$\mu = np$$
$$= 200(0.25)$$
$$= 50.$$

Standard deviation: $\sqrt{np(1 - p)}$
$$= \sqrt{200(0.25)(0.75)}$$
$$= \sqrt{37.5}$$
$$= 6.12 \text{ (rounded)}.$$

Approximate range of values:
$$\mu - 2\sigma < \text{most values} < \mu + 2 \cdot \sigma$$
$$50 - 2(6.12) < \text{most values} < \mu + 2(6.12)$$
$$37.76 < \text{most values} < 62.24.$$

Hence, most values will fall between 38 and 62.

NOTE *The concept of the standard deviation is much more complex than what was presented here. Additional information on the standard deviation will be presented in Chapter 9. More information on the standard deviation can also be found in all statistics textbooks.*

 **PRACTICE**

1. Approximately 10% of people are left handed. If 500 people are randomly selected, find the mean and standard deviation of those who would be left handed.

2. Eighty-five percent of moviegoers purchase something at the concession stand. If 200 moviegoers are randomly selected, find the mean and standard deviation of those who purchase something at the concession stand.

3. A coin is tossed 600 times. Find the mean and standard deviation of the number of heads that will occur. Also, find the approximate range of the number of heads that will occur.

4. If 26 cards are randomly selected from a deck of 52 cards, find the number of face cards (jack, queen, or king) that will be selected on average.

5. If a die is rolled 900 times, find the mean and standard deviation of the number of times a number greater than 4 will occur.

 ANSWERS

1. $n = 500, p = 0.10, 1 - p = 1 - 0.10 = 0.90$

$$\mu = np$$

$$= 500(0.10) = 50$$

$$\sigma = \sqrt{np(1 - p)}$$

$$= \sqrt{500(0.10)(0.90)}$$

$$= \sqrt{45} \approx 6.71.$$

2. $n = 200, p = 0.85, 1 - p = 1 - 0.85 = 0.15$

$$\mu = np$$

$$= 200(0.85) = 170$$

$$\sigma = \sqrt{np(1 - p)}$$

$$= \sqrt{200(0.85)(0.15)}$$

$$= \sqrt{25.5} \approx 5.05$$

3. $n = 600, p = \dfrac{1}{2} = 0.5, 1 - p = 1 - \dfrac{1}{2} = \dfrac{1}{2} = 0.5$

$$\mu = np$$

$$600(0.5) = 300$$

$$\sigma = \sqrt{np\,(1-p)}$$

$$= \sqrt{600(0.5)(0.5)}$$

$$= \sqrt{150} \approx 12.25.$$

$\mu - 2\sigma < \text{most values} < \mu + 2\sigma$

$300 - 2(12.25) < \text{most values} < 300 + 2(12.25)$

$300 - 24.5 < \text{most values} < 300 + 24.5$

$275.5 < \text{most values} < 324.5$ or $276 < \text{most values} < 325$ (rounded).

4. $n = 26; \; p = \dfrac{12}{52} = \dfrac{3}{13}, \; 1 - p = 1 - \dfrac{3}{13} = \dfrac{10}{13}$

$$\mu = np = 26 \cdot \dfrac{3}{13} = 6$$

5. $n = 900; \; p(\text{the number} > 4) = p(5 \text{ or } 6) = \dfrac{2}{6}.$

Hence, $p = \frac{1}{3}$ and $1 - p = \frac{2}{3}$. Then $\mu = np = (900)(\frac{1}{3}) = 300.$

$$\sigma = \sqrt{np(1-p)} = \sqrt{900\left(\dfrac{1}{3}\right)\left(\dfrac{2}{3}\right)}$$

$$= \sqrt{\dfrac{900}{9}(2)} = \sqrt{200}$$

$$\approx 14.14.$$

PROBABILITY SIDELIGHT: Pascal's Triangle

Blaise Pascal (1623–1662) was a French mathematician and philosopher. He made many contributions to mathematics in areas of number theory, geometry, and probability. He is credited along with Fermat for the beginnings of the formal study of probability. He is given credit for developing a triangular array of numbers known as Pascal's triangle, shown here.

```
                 1
              1     1
           1     2     1
        1     3     3     1
     1     4     6     4     1
  1     5    10    10     5     1
1    6    15    20    15     6     1
```

Each number in the triangle is the sum of the number above and to the right of it and the number above and to the left of it. For example, the first 10 in the fifth row is found by adding the 4 and 6 in the fourth row. The first 15 in the sixth row is found by adding the 5 and 10 in the previous row.

The numbers in each row represent the number of different outcomes when coins are tossed. For example, the numbers in row three are 1, 3, 3, and 1. When three coins are tossed, the outcomes are

$$
\begin{array}{cccc}
 & \text{THH} & \text{HTT} & \\
 & \text{HTH} & \text{THT} & \\
\text{HHH} & \text{HHT} & \text{TTH} & \text{TTT}
\end{array}
$$

Notice that there is only one way to get three heads. There are three different ways to get two heads and a tail. There are three different ways to get two tails and a head, and there is one way to get three tails. The same results apply to the genders of the children in a family with three children.

Another property of the triangle is that it represents the answer to the number of combinations of n items taking r items at a time as shown.

$$
\begin{array}{cccc}
 & & {}_0C_0 & & \\
 & {}_1C_0 & & {}_1C_1 & \\
 {}_2C_0 & & {}_2C_1 & & {}_2C_2 \\
{}_3C_0 & & {}_3C_1 & & {}_3C_2 & & {}_3C_3
\end{array}
$$

etc.

The numbers in the triangle have applications in other areas of mathematics such as algebra and graph theory.

It is interesting to note that Pascal included his triangle in a book he wrote in 1653. It wasn't printed until 1665. It is not known if Pascal developed the triangle on his own or heard about it from someone else; however, a similar version of the triangle was found in a Chinese manuscript written by Chi Shi-Kie in 1303!

QUIZ

1. **How many outcomes are there for a binomial experiment?**
   A. 0
   B. 1
   C. 2
   D. It varies

2. **The sum of the probabilities of all outcomes in a probability distribution is**
   A. 0
   B. 1
   C. 2
   D. It varies

3. **Which is not a requirement of a binomial experiment?**
   A. There are two outcomes for each trial.
   B. There is a fixed number of trials.
   C. The outcomes must be dependent.
   D. The probability of a success must be the same for all trials.

4. **The formula for the mean of a binomial distribution is**
   A. np
   B. $np(1-p)$
   C. $n(1-p)$
   D. $\sqrt{np(1-p)}$

5. **The formula for a standard deviation of a binomial distribution is**
   A. np
   B. $np(1-p)$
   C. $n(1-p)$
   D. $\sqrt{np(1-p)}$

6. **If 60% of burglaries occur during the daytime, find the probability that if 12 burglaries occurred last week, five were committed during the day.**
   A. 0.193
   B. 0.899
   C. 0.177
   D. 0.101

7. Forty percent of students in college change their majors at least once. If 20 students are selected, find the probability that 11 have changed their major at least once.

   A. 0.117
   B. 0.176
   C. 0.071
   D. 0.005

8. The state of Kentucky has a smoking rate of 30%. If 800 people are randomly selected, find the mean number of smokers.

   A. 240
   B. 320
   C. 480
   D. 540

9. A survey found that 60% of drivers feel that they are above average drivers. If 1000 drivers are selected, find the number of drivers who feel that they are above average.

   A. 300
   B. 500
   C. 600
   D. 700

10. Twenty percent of M&M candies are yellow. If a sample of 120 M&M candies is selected, find the approximate range of the number that will be yellow. (Round to the nearest whole number.)

   A. 6
   B. 9
   C. 10
   D. 18

SOLUTION

The probabilities are $1, \frac{4}{5}, \frac{3}{5}, \frac{2}{5}$, and $\frac{1}{5}$. The average number of boxes for each are $\frac{1}{4}, \frac{1}{4/5}, \frac{1}{3/5}, \frac{1}{2/5}$, and $\frac{1}{1/5}$; so the total is $1+\frac{5}{4}+\frac{5}{3}+\frac{5}{2}+\frac{5}{1}=11\frac{5}{12}$, which would mean a child on average would need to purchase 12 boxes of cereal since he or she cannot buy $\frac{5}{12}$ of a box.

PRACTICE

1. A die is tossed until a 1, 2, 3, or 4 is obtained. Find the expected number of tosses.

2. On average how many rolls of a die will it take to get four 5s?

3. A card from an ordinary deck of cards is selected and then replaced. Another card is selected, etc. Find the probability that the first heart will occur on the fourth draw.

4. A service station operator gives a scratch-off card with each fill-up over 8 gallons. On each card is one of six colors. When a customer gets all six colors, he wins 10 gallons of gasoline. Find the average number of fill-ups needed to win the 10 gallons.

5. A coin is tossed until six heads are obtained. What is the expected number of tosses?

ANSWERS

1. $p=\frac{4}{6}=\frac{2}{3}$; $\frac{1}{p}=\frac{1}{2/3}=1\div\frac{2}{3}=1\cdot\frac{3}{2}=1.5$ (round up to two, since 1.5 tosses are impossible).

2. $\dfrac{4}{\dfrac{1}{6}}=4\cdot\dfrac{6}{1}=4\cdot6=24.$

3. $\left(\dfrac{3}{4}\right)\left(\dfrac{3}{4}\right)\left(\dfrac{3}{4}\right)\left(\dfrac{1}{4}\right)=\dfrac{27}{256}.$

4. $1+\dfrac{6}{5}+\dfrac{6}{4}+\dfrac{6}{3}+\dfrac{6}{2}+\dfrac{6}{1}=14\dfrac{7}{10}$ or 15 fill-ups.

5. $\dfrac{6}{\left(\dfrac{1}{2}\right)}=6\div\dfrac{1}{2}=6\cdot2=12.$

The Poisson Distribution

Another commonly used discrete distribution is the Poisson distribution (named for Simeon D. Poisson, 1781–1840). This distribution is used when the variable occurs over a period of time, volume, area, etc. For example, it can be used to describe the arrivals of airplanes at an airport, the number of phone calls per hour for a 911 operator, the density of a certain species of plants over a geographic region, or the number of specific bacteria on a fixed surface.

The probability of x successes is

$$\frac{e^{-\lambda}\lambda^x}{x!}$$

where e is a mathematical constant ≈ 2.7183 and λ is the mean or expected value of the variable.

NOTE *The computations require a scientific calculator. Also, tables for values used in the Poisson distribution are available in some statistical textbooks.*

 **EXAMPLE**

A computer help hotline with a toll-free number receives an average of 5 calls per hour. For any given hour, find the probability that it will receive exactly 8 calls. Assume a Poisson distribution.

 SOLUTION

The mean, $\lambda = 5$ and $x = 8$. The probability is

$$\frac{e^{-\lambda}\lambda^x}{x!} = \frac{(2.7183)^{-5} \cdot (5)^8}{8!} = 0.065 \text{ (rounded)}.$$

Hence, there is about a 6.5% probability that the hotline will receive 8 calls.

 **EXAMPLE**

If there are 100 typographical errors randomly distributed in a 600-page manuscript, find the probability that any given page has exactly two errors. Assume a Poisson distribution.

 **SOLUTION**

Find the mean numbers of errors: $\lambda = \frac{150}{600} = \frac{1}{4}$ or 0.25. In other words, there is an average of 0.25 errors per page. In this case, $x = 2$, so the probability of selecting a page with exactly two errors is

$$\frac{e^{-\lambda}\lambda^x}{x!} = \frac{(2.7183)^{-0.25} \cdot (0.25)^2}{2!} = 0.024 \text{ (rounded)}.$$

Hence, the probability of two errors is about 2.4%.

EXAMPLE

Of a mail order company's ads, 1.8% are returned because of the incorrect address or incomplete addresses. If the company sends 500 ads, find the probability that 6 will be returned. Assume a Poisson distribution.

 **SOLUTION**

The average is $\lambda = 0.018 \cdot 500 = 9$ and $x = 6$.

$$\frac{e^{-\lambda}\lambda^x}{x!} = \frac{(2.7183)^{-9} \cdot (9)^6}{6!} = 0.091 \text{ (rounded)}.$$

The probability is about 9.1%.

PRACTICE

1. A company that advertises on the radio in a certain area receives on average 11 calls every time it plays its radio commercials. Find the probability of getting 30 calls if the commercial is aired 3 times a day. Assume a Poisson distribution.

2. In a batch of 50 calculators, on average, two are defective. In a random sample of 100 calculators, find the probability that 10 are defective. Assume a Poisson distribution.

3. If there are three defects in every 1000 ft of cable, find the probability that there would be eight defects in 3000 ft of cable. Assume a Poisson distribution.

4. A bus company finds, on average, that it has to pull out of service five buses per week. Find the probability that for a given week, seven buses will be pulled for repairs. Assume a Poisson distribution.

5. The average number of calls per day to a poison control center is 14. On a given day, find the probability that the center will receive 12 calls.

ANSWERS

1. The average number of calls received is 11 per commercial airing, so if the commercial is aired three times, the average number of calls is $3 \cdot 11 = 33$ calls.

   $\lambda = 33$ and $x = 30$

   $$\frac{e^{-\lambda}\lambda^x}{x!} = \frac{(2.7183)^{-33} \cdot 33^{30}}{30!} = 0.063 \text{ (rounded)}.$$

2. The average number of defective calculators is $\frac{2}{50} = \frac{1}{25}$, so in a batch of 100 calculators, the average would be $100 \cdot \frac{1}{25} = 4$ (rounded).

   $\lambda = 4$ and $x = 10$

   $$\frac{e^{-\lambda}\lambda^x}{x!} = \frac{(2.7183)^{-4} \cdot 4^{10}}{10!} = 0.005 \text{ (rounded)}.$$

3. The average number of defects in 1000 ft of cable is 3, so the average number of defects in 3000 ft of cable is $\frac{3}{1000} \cdot \frac{3000}{1} = 9$.

   $\lambda = 9$ and $x = 8$

   $$\frac{e^{-\lambda}\lambda^x}{x!} = \frac{(2.7183)^{-9} \cdot 9^8}{8!} = 0.132 \text{ (rounded)}.$$

4. In this case, $\lambda = 5$ and $x = 7$

   $$\frac{e^{-\lambda}\lambda^x}{x!} = \frac{(2.7183)^{-5} \cdot 5^7}{7!} = 0.104 \text{ (rounded)}.$$

5. In this case, $\lambda = 14$ and $x = 12$

   $$\frac{e^{-\lambda}\lambda^x}{x!} = \frac{(2.7183)^{-14} \cdot 14^{12}}{12!} = 0.098 \text{ (rounded)}.$$

PROBABILITY SIDELIGHT: Some Thoughts about Lotteries

Today we are bombarded with lotteries. Some people may have the fantasy of winning mega millions for a buck. Each month the prizes seem to be getting larger and larger. Each type of lottery gives the odds and the amount of the winnings. For some lotteries, the amount you win is based on the number of people who play. However, the more people who play the more chance there will be of multiple winners.

The odds and the expected value of a lottery game can be computed using combinations and the probability rules. For example, a lottery game in Pennsylvania is called "Match 6 Lotto." For this game, a player selects 6 numbers from the numbers 1 to 49. If the player matches all six numbers, the player wins a minimum of $500,000. (Note: There are other ways of winning; and if there is no winner, the prize money is held over until the next drawing and increased a certain amount by the number of new players.) For now, just winning the $500,000 will be considered. In order to figure the odds, it is necessary to figure the number of winning combinations. In this case, we are selecting without regard to order six numbers from 49 numbers. Hence, there are $_{49}C_6$ or 13,983,816 ways to select a ticket. However, the odds given in the lottery brochure are 1:4,661,272. The reason is that if you select six numbers, you can have the computer select two more sets of six numbers, giving you three chances to win. So dividing 13,983,816 by 3, you get 4,661,272, and the odds are 1:4,661,272.

Now let's make some sense of this. There are 60 seconds in a minute, 60 minutes in an hour, and 24 hours in a day. So there are $60 \times 60 \times 24 = 86,400$ seconds in a day. If you divide 4,661,272 by 86,400, you get approximately 54 days. So selecting a winner would be like selecting a given random second in a time period of 54 days!

What about a guaranteed method to win the lottery? It does exist. If you purchased all possible number combinations, then you would be assured of winning, wouldn't you? But is it possible?

A group of investors from Melbourne, Australia, decided to try. At the time, they decided to attempt to win a $27 million prize given by the Virginia State Lottery. The lottery consisted of selecting 6 numbers out of a possible 44. This means that they would have to purchase $_{44}C_6$ or 7,059,052 different tickets. At $1 a ticket, they would need to raise $7,059,052. However, the profit would be somewhere near $20 million if they won. Next there is always a possibility of having to split the winnings with other winners, thus reducing the profit. Finally, they need to buy the 7 or so million tickets within the 72-hour time frame. Understand that this group sent out teams, and they were able to purchase about 5 million tickets, and they did win the money without having to split the profit! Some objections were raised by the other players (i.e., losers), but the group was eventually paid off.

By the way, it does not matter which numbers you play on the lottery since the drawing is random and every combination has the same probability of occurring. Some people suggest unusual combinations such as 1, 2, 3, 4, 5, and 6 are better, since there is less of a chance of having to split your winnings if you do indeed win.

So what does this all mean? I heard a mathematician sum it all up by saying that you have the same chance of winning a big jackpot on a state lottery, whether or not you purchase a ticket.

Summary

There are many types of discrete probability distributions besides the binomial distribution. The most common ones are the multinomial distribution, the hypergeometric distribution, the geometric distribution, and the Poisson distribution.

The multinomial distribution is an extension of the binomial distribution and is used when there are three or more independent outcomes for a probability experiment.

The hypergeometric distribution is used when sampling is done without replacement. The geometric distribution is used to determine the probability of an outcome occurring on a specific trial; this distribution can also be used to find the probability of the first occurrence of an outcome.

The Poisson distribution is used when the variable occurs over a period of time, over a period of area or volume, etc.

In addition there are other discrete probability distributions used in mathematics; however, these are beyond the scope of this book.

QUIZ

1. Which distribution can be used when there are three or more outcomes?
   A. Poisson
   B. Hypergeometric
   C. Multinomial
   D. Geometric

2. Which distribution requires that sampling be done without replacement?
   A. Poisson
   B. Geometric
   C. Multinomial
   D. Hypergeometric

3. Which distribution can be used to determine the probability of an outcome occurring on a specific trial?
   A. Geometric
   B. Poisson
   C. Multinomial
   D. Hypergeometric

4. Which distribution can be used when the variable occurs over time?
   A. Poisson
   B. Hypergeometric
   C. Multinomial
   D. Geometric

5. If 4 cards are drawn from a deck without replacement, the probability that exactly 2 hearts will be selected is
   A. 0.035
   B. 0.1875
   C. 0.213
   D. 0.211

6. The probabilities that an income tax return will have 0, 1, 2, or 3 mathematical errors are 0.6, 0.2, 0.15, or 0.05, respectively. If 5 returns are randomly selected, the probability that 2 will contain no errors, 1 will contain one error, one will contain 2 errors, and one will contain 3 errors is
   A. 0.0615
   B. 0.0442
   C. 0.0324
   D. 0.0178

7. A die is rolled five times. The probability of getting two 1s, two 4s, and one 6 is

   A. 0.004
   B. 0.006
   C. 0.013
   D. 0.025

8. A college club consists of 15 women and 12 men. If a committee of six students is selected at random, the probability that exactly two students are men is

   A. 0.0057
   B. 0.3043
   C. 0.2164
   D. 0.0039

9. Of the 12 surfboards in a surf shop, four are white. If five are selected at random to be placed outside on a given day, the probability that exactly two are white is

   A. 0.037
   B. 0.424
   C. 0.22
   D. 0.056

10. About 8% of the cat population carries a certain genetic trait. Assume the distribution is Poisson. The probability that in a group of 200 randomly selected cats, 14 cats carry the gene is

    A. 0.161
    B. 0.093
    C. 0.042
    D. 0.159

11. The number of boating accidents on a large lake follows a Poisson distribution. The probability of an accident is 0.02. If there are 600 boats on the lake on a summer day, the probability that there will be exactly 11 accidents will be

    A. 0.167
    B. 0.248
    C. 0.127
    D. 0.114

12. When a coin is tossed, the probability of getting the first head on the third toss is

    a. $\frac{1}{4}$
    B. $\frac{1}{2}$
    C. $\frac{1}{8}$
    D. $\frac{1}{16}$

13. Eight cards are numbered 1 through 8. The cards are mixed. A card is selected, and its number is recorded. Then it is replaced, and another one is selected and so on. On average how many cards will it take to get all the numbers once?

    A. 28.62
    B. 64
    C. 21.74
    D. 32

14. Cards are selected from a deck of 52 cards and replaced. Find the average number of cards it will take before a face card is selected.

    A. 12
    B. $4\frac{1}{3}$
    C. 8
    D. $\frac{3}{13}$

15. A 10-sided die is rolled; the average number of tosses that it will take to get four 5s is

    A. 30
    B. 8
    C. 18
    D. 40

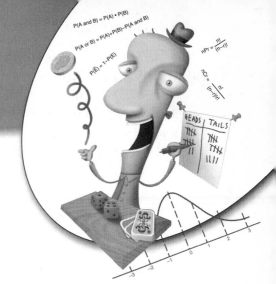

chapter 9

The Normal Distribution

A branch of mathematics that uses probability is called *statistics*. **Statistics** uses observations and measurement to analyze, summarize, make inferences, and draw conclusions based on the various situations of interest to the researcher. This chapter will explain some basic concepts of statistics such as measures of average and measures of variation. Finally, the relationship between probability and the normal distribution will be explained in the last two sections.

CHAPTER OBJECTIVES

In this chapter, you will learn

- How to find the mean, median, and mode for data
- How to find the range and standard deviation for data
- The properties of a normal distribution
- The properties of the standard normal distribution
- How to find the probabilities for a normally distributed variable by transforming it into a standard normal variable

Populations and Samples

Researchers and statisticians use two groups of subjects for studies. They are *populations* and *samples*. A **population** consists of the entire group of subjects under study. Populations must be well defined. For example, if a researcher wanted to do a study on the value of houses in a township, he would have to use all the values for every house in that township. These records could be obtained from the township tax office. The values of the houses are called *data*. **Data** can be measurements such as heights or weights, or observations such as eye color. A list of data values is called a **frequency distribution**.

In many cases, populations contain a large number of subjects, so researchers use a representative subgroup of subjects from the population. This group is called a **sample**. There are several ways to select a sample from a population. These methods can be found in any statistics book and are beyond the scope of this book.

Measures of Average

There are three statistical measures that are commonly used for average. They are the *mean*, *median*, and *mode*. The **mean** is found by adding the data values and dividing the sum by the total number of values. The symbol for a population mean is the Greek letter mu (μ). The symbol for a sample mean is $\overline{X}$.

 EXAMPLE

Find the mean of 32, 30, 45, 18, 24, and 25.

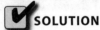 **SOLUTION**

Add the values: 32 + 30 + 45 + 18 + 24 + 25 = 174.

Divide 174 by the number of values, 6; that is, 174 ÷ 6 = 29.

Hence, the mean is 29.

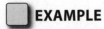 **EXAMPLE**

Ten major earthquakes had magnitudes of 7.0, 6.2, 8.1, 6.8, 5.3, 6.4, 6.8, 7.2, 5.4, and 5.5 as measured on the Richter scale. Find the mean.

SOLUTION

Add: 7.0 + 6.2 + 8.1 + 6.8 + 5.3 + 6.4 + 6.8 + 7.2 + 5.4 + 5.5 = 64.7

Divide by 10; that is, 64.7 ÷ 10 = 6.47.

Hence, the mean of the magnitudes is 6.47.

The **median** is the middle data value, if there is an odd number of data values, or the number halfway between the two data values at the center of the distribution, if there is an even number of data value. The data values must be arranged in order.

EXAMPLE

Find the median of 52, 18, 37, 45, 19, 27, and 14.

SOLUTION

Arrange the data in order: 14, 18, 19, 27, 37, 45, 52.

Find the middle value: 14, 18, 19, <u>27</u>, 37, 45, 52.

The median is 27.

EXAMPLE

Find the median of the number of minutes eight people waited to get service at a local fast-food restaurant when the waiting times were 2, 4, 3, 0, 0, 1, 6, and 5.

SOLUTION

Arrange the data in order: 0, 0, 1, 2, 3, 4, 5, and 6.

The middle of the distribution falls between 2 and 3; hence, the median is (2 + 3) ÷ 2 = 5 ÷ 2 = 2.5.

Hence, the median is 2.5.

The third measure of average is called the *mode*. The **mode** is the data value that occurs most often in a frequency distribution.

EXAMPLE

Find the mode for 12, 17, 42, 17, 15, 19, and 35.

SOLUTION

Since 17 occurs twice and more frequently than any other value, the mode is 17.

EXAMPLE

Find the mode for 9, 10, 12, 12, 12, 15, 18, 20, 20, 20, and 22.

SOLUTION

In this example, 12 and 20 occur most often; hence, 12 and 20 are used as the mode. We say that this distribution is *bimodal*.

EXAMPLE

Find the mode for 9, 16, 37, 22, and 8.

SOLUTION

Since no value occurs more than any other value, there is no mode.

Still Struggling

A distribution can have one mode, more than one mode, or no mode. Also, the mean, median, and mode for a set of values most often differ somewhat from each other. The median and the mode can be found much faster if the data values are first arranged in ascending or descending order.

PRACTICE

1. The number of hamburgers a fast-food restaurant sold each day last week was 837, 642, 591, 1024, 973, 616, and 591. Find the mean, median, and mode for the data.

2. The number of stories in five office buildings in a small city is 18, 15, 9, 10, and 14. Find the mean, median, and mode for the data.

3. The number of automobiles that were parked each day in a small parking lot is 37, 45, 24, 42, 45, 29, 24, and 56. Find the mean, median, and mode for the data.

4. The weights of 11 football players on a starting team are 235, 180, 155, 172, 205, 222, 243, 197, 199, 172, and 214. Find the mean, median, and mode for the data.

5. The number of students in six classrooms of an elementary school is 32, 27, 29, 18, 23, and 25. Find the mean, median, and mode for the data.

 ANSWERS

1. The mean is $\frac{837+642+591+1024+973+616+591}{7} = \frac{5274}{7} = 753.43$ (rounded).

   Arrange entries in ascending order; the median is the underlined entry:

   591, 591, 616, $\underline{642}$, 837, 973, 1024.

   The mode is 591.

2. The mean is $\frac{18+15+9+10+14}{5} = \frac{66}{5} = 13.2$.

   The median is 14.

   There is no mode.

3. The mean is $\frac{37+45+24+42+45+29+24+56}{8} = \frac{302}{8} = 37.75$.

   Arrange entries in ascending order; 24, 24, 29, 37, 42, 45, 45, 56.

   Median $= \frac{37+42}{2} = 39.5$.

   The modes are 24 and 45; the distribution is bimodal.

4. The mean is $\frac{235+180+155+172+205+222+243+197+199+172+214}{11} = \frac{2194}{11} = 199.45$ (rounded).

   Arrange the numbers in ascending order; the median is the under-lined value: 155, 172, 172, 180, 197, $\underline{199}$, 205, 214, 222, 235, 243.

   Median $= 199$.

   The mode is 172.

5. The mean $= \frac{32+27+29+18+23+25}{6} = \frac{154}{6} = 25.67$ (rounded).

   In ascending order, the entries are 18, 23, 25, 27, 29, 32.

   Median $= \frac{25+27}{2} = \frac{52}{2} = 26$.

   There is no mode.

Measures of Variability

In addition to measures of average, statisticians are interested in measures of variability. One measure of variability is called the *range*. The **range** is the difference between the largest data value and the smallest data value.

 EXAMPLE

Find the range for 18, 13, 25, 32, 15, and 27.

 SOLUTION

Since the largest data value is 32 and the smallest data value is 13, the range is 32 − 13 = 19. In statistics, the range is always a single number. Do not say the range is 13 to 32.

Another measure that is also used as a measure of variability for data is called the **standard deviation**. This measure was also used in Chapter 7. The symbol for the population deviation is the lower case Greek letter sigma (σ). The symbol for the sample standard deviation is *s*.

The steps for computing the standard deviation for individual data values are

Step 1: Find the mean.

Step 2: Subtract the mean from each value and square the differences.

Step 3: Find the sum of the squares.

Step 4: Divide the sum by the number of data values minus 1.

Step 5: Take the square root of the answer.

 EXAMPLE

Find the standard deviation for 17, 24, 16, 26, and 17.

 SOLUTION

Step 1: Find the mean: $\frac{17+24+16+26+17}{5} = \frac{100}{5} = 20$.

Step 2: Subtract the mean from each value and square the differences:

$$17 - 20 = -3 \qquad (-3)^2 = 9$$
$$24 - 20 = 4 \qquad 4^2 = 16$$
$$16 - 20 = -4 \qquad (-4)^2 = 16$$
$$26 - 20 = 6 \qquad 6^2 = 36$$
$$17 - 20 = -3 \qquad (-3)^2 = 9.$$

Step 3: Find the sum of the squares:
$$9 + 16 + 16 + 36 + 9 = 86.$$
Step 4: Divide 86 by 5 − 1 or 4: 86 ÷ 4 = 21.5.
Step 5: Take the square root of the answer $\sqrt{21.5} = 4.64$ (rounded).
The standard deviation is 4.64.

Recall from Chapter 7 that most data values fall within 2 standard deviations of the mean. In this case, $20 \pm 2 \cdot (4.64)$ is 20 ± 9.28 or $10.72 <$ most values < 29.28. Looking at the data, you can see all the data values fall between 10.72 and 29.28.

 EXAMPLE

The number of miles that 10 people ran in one day is 5, 8, 3, 6, 5, 2, 1, 3, 1, and 6. Find the standard deviation for the data.

SOLUTION

Step 1: Find the mean: $\frac{5+8+3+6+5+2+1+3+1+6}{10} = \frac{40}{10} = 4.$

Step 2: Subtract and square:

$5 - 4 = 1$	$1^2 = 1$
$8 - 4 = 4$	$4^2 = 16$
$3 - 4 = -1$	$(-1)^2 = 1$
$6 - 4 = 2$	$2^2 = 4$
$5 - 4 = 1$	$1^2 = 1$
$2 - 4 = -2$	$(-2)^2 = 4$
$1 - 4 = -3$	$(-3)^2 = 9$
$3 - 4 = -1$	$(-1)^2 = 1$
$1 - 4 = -3$	$(-3)^2 = 9$
$6 - 4 = 2$	$2^2 = 4.$

Step 3: Find the sum: $1 + 16 + 1 + 4 + 1 + 4 + 9 + 1 + 9 + 4 = 50.$

Step 4: Divide by 9: $50 \div 9 = 5.56$

Step 5: Take the square root: $\sqrt{5.56}$ (rounded).

The standard deviation is 2.34.

Still Struggling

Basically, the standard deviation is the square root of the "average" of the squared distances, where a distance equals the difference between a data value and the mean. The reason that we square the differences is that the sum of the differences without squaring will always be 0. We divide by $n - 1$ instead of n to get a better approximation of the population standard deviation when the sample size is small.

PRACTICE

1. The ages of 9 of the wealthiest people in the United States are 81, 78, 61, 43, 68, 57, 40, 71, and 41. Find the range and standard deviation for the data.

2. The number of emergency patients a hospital has seen for a six-hour period is 27, 32, 15, 19, 44, and 25. Find the range and standard deviation for the data.

3. The number of miles per gallon 7 randomly selected new cars obtained is, respectively, 32, 12, 18, 19, 18, 27, and 28. Find the range and standard deviation for the data.

4. The gasoline tax (in cents) per gallon for five selected states is 2, 17, 30, 20, and 16. Find the range and standard deviation for the data.

5. The weight gain (in pounds) over a year for 9 people is 6, 8, 3, 2, 5, 15, 3, 4, and 8. Find the range and standard deviation for the data.

ANSWERS

1. The range is $81 - 40 = 41$.

   The mean is $\frac{81+78+61+43+68+57+40+71+41}{9} = \frac{540}{9} = 60$.

   The standard deviation is found as follows:

$81 - 60 = 21$	$21^2 = 441$
$78 - 60 = 18$	$18^2 = 324$
$61 - 60 = 1$	$1^2 = 1$
$43 - 60 = -17$	$(-17^2) = 289$
$68 - 60 = 8$	$8^2 = 64$

$$57 - 60 = -3 \qquad (-3)^2 = \quad 9$$
$$40 - 60 = -20 \qquad (-20)^2 = \quad 400$$
$$71 - 60 = 11 \qquad 11^2 = \quad 121$$
$$41 - 60 = -19 \qquad (-19)^2 = \underline{\quad 361}$$
$$2010$$

$$\frac{2010}{8} = 251.25$$

$\sqrt{251.25} = 15.851$ (rounded) = standard deviation.

2. The range is $44 - 15 = 29$.

   The mean is $\frac{27+32+15+19+44+25}{6} = \frac{162}{6} = 27$.

   The standard deviation is calculated as follows:

$$27 - 27 = 0 \qquad 0^2 = \quad 0$$
$$32 - 27 = 5 \qquad 5^2 = \quad 25$$
$$15 - 27 = -12 \qquad (-12)^2 = 144$$
$$19 - 27 = -8 \qquad (-8)^2 = \quad 64$$
$$44 - 27 = 17 \qquad 17^2 = 289$$
$$25 - 27 = -2 \qquad (-2)^2 = \underline{\quad 4}$$
$$526$$

$$\frac{526}{5} = 105.2$$

   The standard deviation = $\sqrt{105.2} = 10.26$ (rounded).

3. The range is $32 - 12 = 20$.

   The mean is $\frac{32+12+18+19+18+27+28}{7} = \frac{154}{7} = 22$.

   The standard deviation is found by

$$32 - 22 = 10 \qquad 10^2 = 100$$
$$12 - 22 = -10 \qquad (-10)^2 = 100$$
$$18 - 22 = -4 \qquad (-4)^2 = \quad 16$$
$$19 - 22 = -3 \qquad (-3)^2 = \quad 9$$
$$18 - 22 = -4 \qquad (-4)^2 = \quad 16$$
$$27 - 22 = 5 \qquad 5^2 = \quad 25$$
$$28 - 22 = 6 \qquad 6^2 = \underline{\quad 36}$$
$$302$$

$$\frac{302}{6} = 50.33 \text{ (rounded)}$$

   The standard deviation is $\sqrt{50.33} = 7.09$ (rounded).

4. The range is 30 – 2 = 28.

    The mean is $\frac{2+17+30+20+16}{5} = \frac{85}{5} = 17$.

    To find the standard deviation:

2 – 17 = –15	$(-15)^2 = 225$
17 – 17 = 0	$0^2 = 0$
30 – 17 = 13	$13^2 = 169$
20 – 17 = 3	$3^2 = 9$
16 – 17 = –1	$(-1)^2 = \underline{1}$
	404

    Standard deviation $= \sqrt{\frac{404}{5-1}} = \sqrt{101} = 10.05$ (rounded).

5. The range is 15 – 2 = 13.

    The mean is $\frac{6+8+3+2+5+15+3+4+8}{9} = \frac{54}{9} = 6$.

    The standard deviation is found by

6 – 6 = 0	$0^2 = 0$
8 – 6 = 2	$2^2 = 4$
3 – 6 = –3	$(-3)^2 = 9$
2 – 6 = –4	$(-4)^2 = 16$
5 – 6 = –1	$(-1)^2 = 1$
15 – 6 = 9	$9^2 = 81$
3 – 6 = –3	$(-3)^2 = 9$
4 – 6 = –2	$(-2)^2 = 4$
8 – 6 = 2	$2^2 = \underline{4}$
	128

    $$\frac{128}{8} = 16$$

    The standard deviation is $\sqrt{16} = 4$.

The Normal Distribution

Recall from Chapter 7 that a continuous random variable can assume all values between any two given values. For example, the heights of adult males is a continuous random variable since a person's height can be any reasonable number.

We are, however, limited by our measuring instruments. The variable temperature is a continuous variable since temperature can assume any numerical value between any two given values. Many continuous variables can be represented by formulas and graphs or curves. These curves represent probability distributions. In order to find probabilities for values of a variable, the area under the curve between two given values is used.

One of the most often used continuous probability distributions is called the **normal probability distribution**. Many variables are approximately normally distributed and can be represented by the normal distribution. It is important to realize that the normal distribution is a perfect theoretical mathematical curve, but no real-life variable is perfectly normally distributed.

The real-life normally distributed variables can be described by the theoretical normal distribution. This is not so unusual when you think about it. Consider the wheel. It can be represented by the mathematically perfect circle, but no real-life wheel is perfectly round. The mathematics of the circle, then, is used to describe the wheel.

The normal distribution has the following properties:

1. It is bell shaped.

2. The mean, median, and mode are at the center of the distribution.

3. It is symmetric about the mean. (This means that it is a reflection of itself when the mean is at the center.)

4. It is continuous; that is, there are no gaps.

5. It never touches the x-axis.

6. The total area under the curve is 1.00 or 100%.

7. About 0.68 or 68% of the area under the curve falls within one standard deviation on either side of the mean. (Recall that μ is the symbol for the mean and σ is the symbol for the standard deviation.) About 0.95 or 95% of the area under the curve falls within two standard deviations of the mean. About 1.00 or 100% of the area falls within three standard deviations of the mean.

NOTE *It is somewhat less than 100%, but for simplification, 100% will be used here.*

See Figure 9-1. The decimals indicate the proportion of the area between the two parameters. For example, the area between the mean, μ, and 1 standard deviation above the mean, $\mu + 1\sigma$, is 34.1% of the total area under the curve.

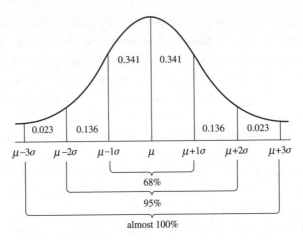

FIGURE 9-1

We can use this information to solve problems when a variable is normally distributed.

EXAMPLE

The mean systolic blood pressure of a group of people is 120. The standard deviation is 6. Assume the variable is normally distributed. Find the probability that a person's blood pressure is between 120 and 132.

SOLUTION

Draw the normal distribution and place the mean, 120, at the center. Then place the mean +1 standard deviation (126) to the right, the mean +2 standard deviations (132) to the right, the mean +3 standard deviations (138) to the right, the mean –1 standard deviation (114) to the left, the mean –2 standard deviations (108) to the left, and the mean –3 standard deviations (102) to the left, as shown in Figure 9-2.

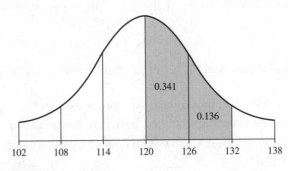

FIGURE 9-2

Using the areas shown in Figure 9-1, the area under the curve between 120 and 132 is 0.341 + 0.136 = 0.477 or 47.7%. Hence, the probability that a person's systolic blood pressure is between 120 and 132 is about 48%.

EXAMPLE

The average waiting time to see a medical doctor in the emergency room is 18 minutes. The standard deviation is 4 minutes. If a person goes to the emergency room, find the probability that he or she will wait between 6 and 14 minutes to see a doctor. Assume the variable is normally distributed.

SOLUTION

Draw the normal distribution curve and place 18 at the center; then place 22, 26, and 30 to the right corresponding to 1, 2, and 3 standard deviations above the mean, and 14, 10, and 6 to the left corresponding to 1, 2, and 3 standard deviations below the mean. Now place the areas (percents) on the graph. See Figure 9-3.

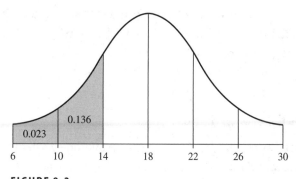

FIGURE 9-3

Since we are finding the probability for the number of minutes between 6 and 14, add the areas 0.136 + 0.023 = 0.159 or 15.9%. Hence, the probability is about 16%.

EXAMPLE

The average size of an American diamond engagement ring is 0.9 carats. If the standard deviation is 0.3 carats, find the probability that a randomly selected diamond engagement ring is greater than 0.6 carats. Assume the variable is normally distributed.

SOLUTION

Draw the normal distribution curve and place 0.9 at the center. Place 1.2, 1.5, and 1.8 to the right and 0.6, 0.3, and 0 to the left, corresponding to 1, 2, and 3 standard deviations above and below the mean, respectively. Fill in the corresponding areas. See Figure 9-4.

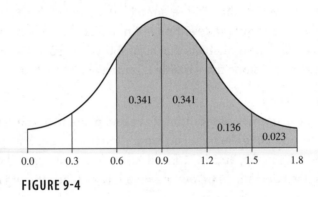

FIGURE 9-4

Since we are interested in any size greater than 0.6, add 0.341 + 0.341 + 0.136 + 0.023 = 0.841 = 84.1%. Hence, with an 84% probability, a randomly selected ring is more than 0.6 carats.

PRACTICE

1. It has been found that families recycle on average 28 lb of newspapers per year. If the standard deviation is 3, find the probability that a family recycles between 28 lb and 34 lb of newspapers per year. Assume the variable is normally distributed.

2. It takes on average 16 minutes for an emergency service to respond to calls. The standard deviation is 4.5 minutes. Find the probability that it will take the service between 7 minutes and 20.5 minutes to respond to a call. Assume the variable is normally distributed.

3. If it takes a courier 200 minutes to travel between Pittsburgh and Harrisburg, find the probability that it will take the courier less than 190 minutes to make the trip. The standard deviation is 10 minutes. Assume the variable is normally distributed.

4. The average amount of snow in Springville is 49 inches per year. The standard deviation is 5 inches. Find the probability that Springville will receive more than 59 inches in 1 year. Assume the variable is normally distributed.

5. A study was done to find the average waiting time to get service in a local bank. The mean waiting time was 9.2 minutes, and the standard deviation was 2.6 minutes. If a person enters the bank, find the probability that the person has to wait between 6.6 minutes and 11.8 minutes.

ANSWERS

1. The required area is shown in Figure 9-5.

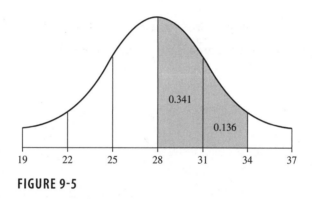

FIGURE 9-5

Probability = 0.341 + 0.136 = 0.477 or 47.7%.

2. The required area is shown in Figure 9-6.

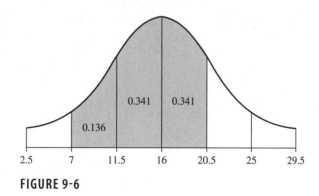

FIGURE 9-6

Probability = 0.136 + 0.341 + 0.341 = 0.818 or 81.8%.

3. The required area is shown in Figure 9-7.

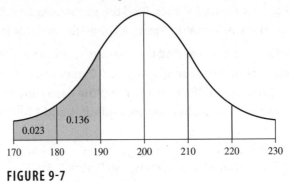

FIGURE 9-7

Probability = 0.023 + 0.136 = 0.159 or 15.9%.

4. The required area is shown in Figure 9-8.

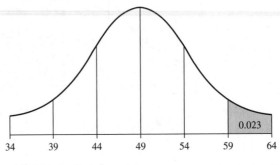

FIGURE 9-8

Probability = 0.023 or 2.3%.

5. The required area is shown in Figure 9-9.

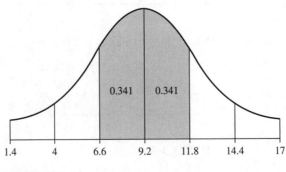

FIGURE 9-9

Probability = 0.341 + 0.341 = 0.682 = 68.2%.

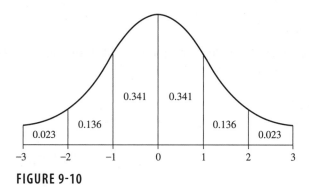

FIGURE 9-10

The Standard Normal Distribution

The normal distribution can be used as a model to solve many problems about variables that are approximately normally distributed. Since each variable has its own mean and standard deviation, statisticians use what is called the *standard normal distribution* to solve the problems.

The **standard normal distribution** has all the properties of a normal distribution, but the mean is 0 and the standard deviation is 1. See Figure 9-10.

A value for any variable that is approximately normally distributed can be transformed into a standard normal value by using the following formula:

$$z = \frac{\text{value} - \text{mean}}{\text{standard deviation}} \quad \text{or} \quad \frac{x - \mu}{\sigma}$$

The standard normal values are called **z values** or **z scores**.

 **EXAMPLE**

Find the corresponding *z* value for a value of 17 if the mean of a variable is 10 and the standard deviation is 4.

SOLUTION

$$z = \frac{\text{value} - \text{mean}}{\text{standard deviation}} = \frac{17 - 10}{4} = \frac{7}{4} = 1.75$$

Hence, the *z* value of 1.75 corresponds to a value of 17 for an approximately normal distribution which has a mean of 10 and a standard deviation of 4. The *z* values are negative for values of variables that are below the mean.

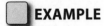

EXAMPLE

Find the corresponding *z* value for a value of 14 if the mean of a variable is 20 and the standard deviation is 5.

SOLUTION

$$z = \frac{\text{value} - \text{mean}}{\text{standard deviation}} = \frac{14 - 20}{5} = -\frac{6}{5} = -1.2.$$

Hence, in this case a value of 14 is equivalent to a *z* value of –1.2.

In addition to finding probabilities for values that are between 0, 1, 2, and 3 standard deviations of the mean, probabilities for other values can be found by converting them to *z* values and using the standard normal distribution.

Areas between any two given *z* values under the standard normal distribution curve can be found by using calculus; however, tables for specific *z* values can be found in any statistics textbook. An abbreviated table of areas is shown in Table 9-1.

TABLE 9-1	Approximate Cumulative Areas for the Standard Normal Distribution*						
z	**Area**	**z**	**Area**	**z**	**Area**	**z**	**Area**
−3.0	.001	−1.5	.067	0.0	.500	1.5	.933
−2.9	.002	−1.4	.081	0.1	.540	1.6	.945
−2.8	.003	−1.3	.097	0.2	.579	1.7	.955
−2.7	.004	−1.2	.115	0.3	.618	1.8	.964
−2.6	.005	−1.1	.136	0.4	.655	1.9	.971
−2.5	.006	−1.0	.159	0.5	.692	2.0	.977
−2.4	.008	−0.9	.184	0.6	.726	2.1	.982
−2.3	.011	−0.8	.212	0.7	.758	2.2	.986
−2.2	.014	−0.7	.242	0.8	.788	2.3	.989
−2.1	.018	−0.6	.274	0.9	.816	2.4	.992
−2.0	.023	−0.5	.309	1.0	.841	2.5	.994
−1.9	.029	−0.4	.345	1.1	.864	2.6	.995
−1.8	.036	−0.3	.382	1.2	.885	2.7	.997
−1.7	.045	−0.2	.421	1.3	.903	2.8	.997
−1.6	.055	−0.1	.460	1.4	.919	2.9	.998
						3.0	.999

*By "Cumulative Area" here is meant the area under the curve, from the extreme left to the *z* value in question.

This table gives the approximate cumulative areas for z values between –3 and +3. The next three examples will show how to find the area (and corresponding probability in decimal form).

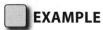 **EXAMPLE**

Find the area under the standard normal distribution curve to the left of $z = 1.6$.

 **SOLUTION**

The area is shown in Figure 9-11.

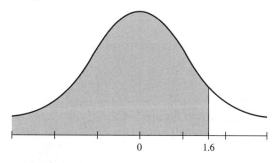

0   1.6

FIGURE 9-11

In order to find the area under the standard normal distribution curve to the left of any given z value, just look it up directly in Table 9-1. The area is 0.945 or 94.5%.

 **EXAMPLE**

Find the area under the standard normal distribution curve between $z = -2.1$ and $z = 0.3$.

 **SOLUTION**

The area is shown in Figure 9 -12.

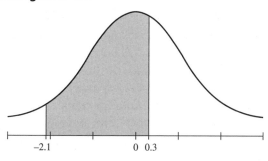

–2.1   0 0.3

FIGURE 9-12

To find the area under the standard normal distribution curve between any given two z values, look up the areas in Table 9-1 and subtract the smaller area from the larger. In this case, the area corresponding to $z = -2.1$ is 0.018, and the area corresponding to $z = 0.3$ is 0.618, so the area between $z = -2.1$ and $z = 0.3$ is $0.618 - 0.018 = 0.6 = 60\%$. In other words, 60% of the area under the standard normal distribution curve is between $z = -2.1$ and $z = 0.3$.

 EXAMPLE

Find the area under the standard normal distribution curve to the right of $z = -1.9$.

 SOLUTION

The area is shown in Figure 9-13.

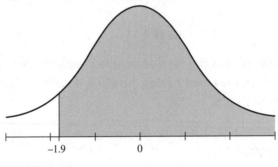

$$-1.9 \qquad 0$$

FIGURE 9-13

To find the area under the standard normal distribution curve to the right of any given z value, look up the area in the table and subtract that from 1 since the entire area under the curve is 1. The area to the left of $z = -1.9$ is 0.029. Hence, $1 - 0.029 = 0.971$. In other words, 0.971 of the area under the standard normal distribution curve lies to the right of $z = -1.9$.

Using Table 9-1 and the formula for transforming values for variables that are approximately normally distributed, you can find the probabilities of various events. Recall that μ = the mean and σ = the standard deviation. Use the formula $z = \frac{x - \mu}{\sigma}$.

**EXAMPLE**

The scores on a certain test are normally distributed and have a mean of 80 and a standard deviation of 8. If a student who took the test is selected at random, find the probability that the student scored above 82.

**SOLUTION**

Draw a normal distribution and shade the desired area. See Figure 9-14.

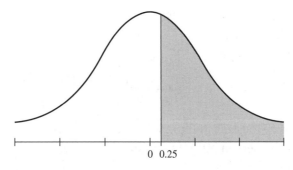

0   0.25

FIGURE 9-14

Find the *z* value. (Round it to one decimal place if necessary.)

$$z = \frac{x - \mu}{\sigma} = \frac{82 - 80}{8} = \frac{2}{8} = 0.25$$

Round the value to 0.3. Look up the area in Table 9-1 that corresponds to 0.3. It is 0.618.

Since the area to the right of 82 is desired, subtract 1 − 0.618 = 0.382. Hence, the probability that a student scores above 82 is 0.382 or 38.2%.

**EXAMPLE**

A brisk walk at 4.5 miles per hour burns an average of 300 calories. The variable is normally distributed, and the standard deviation is 10 calories. If a person walks at 4.5 miles per hour, find the probability that she burns less than 275 calories.

SOLUTION

Draw the normal distribution curve and shade the desired area. See Figure 9-15.

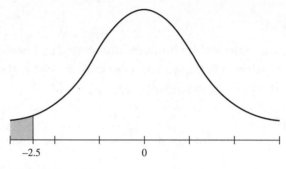

FIGURE 9-15

Find the *z* value:

$$z = \frac{x - \mu}{\sigma} = \frac{275 - 300}{10} = -2.5$$

Look up the area in Table 9-1 for $z = -2.5$. It is 0.006. Hence, the probability that the person burns less than 275 calories is 0.006 or 0.6%.

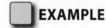 **EXAMPLE** _____

In a study it was found that people keep their televisions for an average of 4.8 years. The standard deviation is 1 year, and the variable is normally distributed. If a television is selected at random, find the probability that it is between 5 and 6 years old.

**SOLUTION** _____

Draw the standard normal distribution and show the desired area. See Figure 9-16.

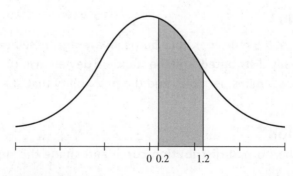

FIGURE 9-16

Find the z values:

$$z = \frac{x - \mu}{\sigma} = \frac{5 - 4.8}{1} = 0.2$$

$$z = \frac{x - \mu}{\sigma} = \frac{6 - 4.8}{1} = 1.2$$

Find the two areas corresponding to 0.2 and 1.2. For $z = 0.2$, the area is 0.579. For $z = 1.2$, the area is 0.885. Subtract the areas: $0.885 - 0.579 = 0.306$ or 30.6%. Hence, the probability that a randomly selected television set is between 5 and 6 years old is 30.6%.

PRACTICE

1. For a specific model of automobile, the average gasoline consumption is 28 miles per gallon. The standard deviation is 3 miles, and the variable is normally distributed. If an automobile is selected at random, find the probability that it will get less than 25 miles per gallon.

2. The average salary for production workers is $12.00 per hour. The standard deviation is $2.00, and the variable is normally distributed. If a worker is selected at random, find the probability that he or she makes between $10.75 and $12.25.

3. The average temperature of Laurel Lake in September is 64°. The standard deviation is 1.5° and the variable is normally distributed. If a day in September is selected at random, find the probability that the temperature is greater than 65°.

4. The average time it takes an army recruit to run an obstacle course is 18 minutes. The standard deviation is 4 minutes, and the variable is normally distributed. If a recruit is selected at random, find the probability that he or she will run the course in less than 15 minutes.

5. The mean of the serum cholesterol levels of a certain group of individuals is 206. The standard deviation is 6. If a person is selected from the group at random, find the probability that his or her cholesterol level is between 203 and 218. Assume the variable is normally distributed.

 ANSWERS

1. $z = \dfrac{25 - 28}{3} = \dfrac{-3}{3} = -1.$

    The required area is shown in Figure 9-17. The area that corresponds to a $z = -1$ is 0.159. Since we are looking for the area to the left of $z = -1$,

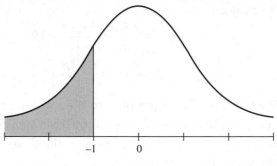

FIGURE 9-17

we use 0.159, and its corresponding percent is 15.9%. Hence, the probability that an automobile gets less than 25 miles per gallon is 15.9%. See Figure 9-17.

2. $z = \dfrac{10.75 - 12.00}{2.00} = -0.625$ or -0.6 (rounded).

The area corresponding to $z = -0.6$ is 0.274.

$$z = \dfrac{12.25 - 12.00}{2.00} = 0.125 \text{ or } 0.1 \text{ (rounded)}.$$

The area in Tale 9-1 corresponding to $z = 0.1$ is 0.540. Thus, the desired area = area between the above z values = 0.540 − 0.274 = 0.266 = 26.6% = shaded section in Figure 9-18.

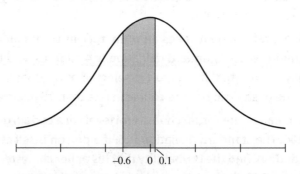

FIGURE 9-18

3. $z = \dfrac{65 - 64}{1.5} = 0.667$ or 0.7 (rounded).

The area in Table 9-1 corresponding to $z = 0.7$ is 0.758. Since we are looking for the area to the right of $z = 0.7$, we subtract 1 − 0.758 = 0.242.

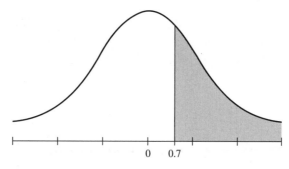

FIGURE 9-19

See Figure 9-19. Hence, the probability that the temperature is greater than 65° is 24.2%.

4. $z = \dfrac{15-18}{4} = -\dfrac{3}{4} = -0.75$ or -0.8 (rounded).

From Table 9-1, the area for $z = -0.8$ is 0.212. Since we want the probability that the course run is less than 15 minutes, we just take the said area, and have 21.2% as our answer. See Figure 9-20.

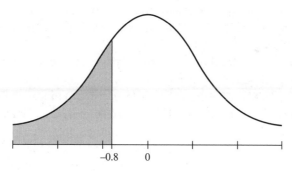

FIGURE 9-20

5. $z = \dfrac{203-206}{6} = -\dfrac{3}{6} = -0.5.$

The area from Table 9-1 corresponding to $z = -0.5$ is 0.309.

$$z = \frac{218-206}{6} = -\frac{12}{6} = 2$$

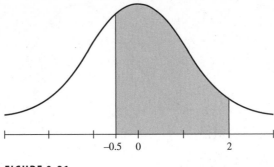

FIGURE 9-21

The area from Table 9-1 corresponding to $z = 2$ is 0.977. Since we are looking for the area between 203 and 218, we subtract the two areas, 0.977 − 0.309 = 0.668 or 66.8%. See Figure 9-21.

PROBABILITY SIDELIGHT: A Brief History of the Normal Distribution

The applications of the normal distribution are many and varied. It is used in astronomy, biology, business, education, medicine, engineering, psychology, and many other areas. The development of the concepts of the normal distribution is quite interesting.

It is believed that the first mathematician to discover some of the concepts associated with the normal distribution was the English mathematician, Abraham De Moivre (1667–1754). He was born in France, but moved to England because of the French government's restrictions on religion and civil liberties. He supported himself by becoming a private tutor in mathematics. He studied the probability of tossing coins, rolling dice, and other forms of gambling. In 1716, he wrote a book on gambling entitled *The Doctrine of Chances.* In addition to his tutoring, wealthy patrons came to him to find out what the payoff amount should be for various gambling games. He made many contributions to mathematics. In probability, he tossed a large number of coins many times and recorded the number of heads that resulted on each trial. He found that approximately 68% of the results fell within a predictable distance (now called the standard deviation) on either side of the mean and that 95% of the results fell within two predictable distances on either side of the mean. In addition, he noticed that the shape of the distribution was bell shaped, and he derived the equation for the normal curve in 1733, but his work in this area of mathematics went relatively unnoticed for a long period of time.

In 1781, a French mathematician, Pierre Simon Laplace (1749–1827), was studying the gender of infants, attempting to prove that the number of males born was slightly more than the number of females born. (This fact has been verified today.) Laplace noticed that the distribution of births was also bell shaped and that the outcomes followed a particular pattern. Laplace also developed a formula for the normal distribution, and it is thought that he was unaware of DeMoivre's earlier work.

About 30 years later in 1809, a German mathematician, Carl Friedrich Gauss (1777–1855), deduced that the errors in the measurements of the planets due to imperfections in the lenses in telescopes and the human eye were approximately bell shaped. The theory was called Gauss' Law of Error. Gauss developed a complex measure of variation for the data and also an equation for the normal distribution curve. The curve is sometimes called the Gaussian distribution in his honor. In addition to mathematics, Gauss also made many contributions to astronomy.

During the 1800s, at least seven different measures of variation were used to describe distributions. It wasn't until 1893 that the statistician Karl Pearson coined the term "standard deviation."

Around 1830, researchers began to notice that the normal distribution could be used to describe other phenomena. For example, in 1846, Adolphe Quetelet (1796–1874) began to measure the chest sizes of Scottish soldiers. He was trying to develop the concept of the "average man," and found the normal distribution curve was applicable to these measurements. Incidentally, Quetelet also developed the concept of body mass index, which is still used today.

A German experimental psychologist, Hermann Ebbinghaus (1855–1913), found that the normal distribution was applicable to measures of intelligence and memorization in humans.

Sir Francis Galton in 1889 invented a device that he called a "Quincunx." This device dropped beads through a series of pegs into slots whose heights resulted in a bell-shaped graph. It wasn't until 1924 that Karl Pearson found that DeMoivre had discovered the formula for the normal distribution curve long before Laplace or Gauss.

Summary

Statistics is a branch of mathematics that uses probability. Statisticians use data to analyze, summarize, make inferences, and draw conclusions from data. They obtain data from populations or samples. A population consists of all subjects under study. A sample is a representative subgroup of subjects taken from a population.

There are three commonly used measures of average: they are the mean, median, and mode. The mean is the sum of the data values divided by the number of data values. The median is the midpoint of the data values when they are arranged in numerical order. The mode is the data value that occurs most often.

There are two commonly used measures of variability. They are the range and standard deviation. The range is the difference between the smallest data value and the largest data value. The standard deviation is the square root of the average of the squares of the differences each value is from the mean.

Many variables are approximately normally distributed and the standard normal distribution can be used to find probabilities for various situations involving values of these variables.

The standard normal distribution is a continuous, bell-shaped curve such that the mean, median, and mode are at its center. It is also symmetrical about the mean. The mean is equal to 0 and the standard deviation is equal to 1. About 68% of the area under the standard normal distribution lies within one standard deviation of the mean, about 95% within 2 standard deviations, and about 100% within 3 standard deviations of the mean.

QUIZ

1. **What is the mean of 32, 27, 45, 63, 41, and 50?**
   A. 36
   B. 43
   C. 44
   D. 50

2. **What is the median of 15, 19, 33, 27, 22, and 46?**
   A. 15.5
   B. 20
   C. 24.5
   D. 31

3. **What is the mode of 27, 56, 19, 43, 55, 58, and 56?**
   A. 56
   B. 44
   C. 55
   D. None

4. **What is the range of 97, 101, 92, 105, and 94?**
   A. 13
   B. 5
   C. 10
   D. 3

5. **What is the standard deviation of 19, 24, 16, 32, and 9? (Round to 2 decimal spaces.)**
   A. 3.9
   B. 18.8
   C. 8.63
   D. 17.4

6. **The area under the standard normal distribution is**
   A. Unknown
   B. 1 or 100%
   C. Infinite
   D. 0.5 or 50%

7. **For the standard normal distribution,**
   A. The mean = 1 and the standard deviation = 2.
   B. The mean = 0 and the standard deviation = 1.
   C. The mean and the standard deviation can both vary.
   D. The mean = 1 and the standard deviation = 0.

8. Which is **not** a property of the normal distribution?
   A. It is not symmetrical about the mean.
   B. The mean is at the center.
   C. It is continuous.
   D. It is bell shaped.

9. When the value of a variable is transformed into a standard normal variable, the new value is called a(n)
   A. y value
   B. 0 value
   C. z value
   D. x value

10. What percent of the area under the standard normal distribution lies within 2 standard deviations of the mean?
    A. 100%
    B. 95%
    C. 68%
    D. Unknown

Use Table 9-1 to answer Questions 11 through 15.

11. The area under the standard normal distribution to the right of $z = 1.6$ is
    A. 5.5%
    B. 94.5%
    C. 62.5%
    D. 37.5 %

12. The area under the standard normal distribution to the left of $z = -2.1$ is
    A. 96.4%
    B. 37.5%
    C. 98.2%
    D. 1.8%

13. The area under the standard normal distribution between $z = -1.4$ and $z = 0.9$ is
    A. 81.6%
    B. 89.7%
    C. 73.5%
    D. 8.1%

14. An exam which is approximately normally distributed has a mean of 200 and a standard deviation of 20. If a person who took the exam is selected at random, find the probability that the person scored above 216.

    A. 78.8%
    B. 63.2%
    C. 6.7%
    D. 21.2%

15. The average height for adult females is 64 in. Assume the variable is normally distributed with a standard deviation of 2 in. If a female is randomly selected, find the probability that her height is between 66 and 68 in.

    A. 13.6%
    B. 97.7%
    C. 81.8%
    D. 84.1%

chapter **10**

Simulation

Instead of studying actual situations that sometimes might be too costly, too dangerous, or too time consuming, researchers create similar situations using random devices so that they are less expensive, less dangerous, or less time consuming. For example, pilots use flight simulators to practice on before they actually fly a real plane. Many video games use the computer to simulate real-life sports situations such as baseball, football, or hockey.

Simulation techniques go back to ancient times when the game of chess was invented to simulate warfare. Modern techniques date to the mid-1940s when two physicists, John Von Neuman and Stanislaw Ulam, developed simulation techniques to study the behavior of neutrons in the design of atomic reactors.

Mathematical simulation techniques use random number devices along with probability to create conditions similar to those found in real life. Random devices are items such as dice, coins, and computers or calculators. These devices generate what are called *random numbers*. For example, when a fair die is rolled, it generates the numbers 1 through 6 randomly. This means that the outcomes occur by chance and each outcome has the same probability of occurring.

Computers play an important role in simulation since they can generate random numbers, perform experiments, and tally the results much faster than humans can. In this chapter, the concepts of simulation will be explained by using dice or coins.

CHAPTER OBJECTIVE

In this chapter, you will learn

- How to find the probability of an event by using simulation techniques

The Monte Carlo Method

The Monte Carlo Method of simulation uses random numbers. The steps are

Step 1: List all possible outcomes of the experiment.

Step 2: Determine the probability of each outcome.

Step 3: Set up a correspondence between the outcomes of the experiment and random numbers.

Step 4: Generate the random numbers (i.e., roll the dice, toss the coin, etc.)

Step 5: Repeat the experiment and tally the outcomes.

Step 6: Compute any statistics and state the conclusions.

If an experiment involves two outcomes and each has a probability of $\frac{1}{2}$, a coin can be tossed. A head would represent one outcome and a tail the other outcome. If a die is rolled, an even number could represent one outcome and an odd number could represent the other outcome. If an experiment involves 5 outcomes, each with a probability of $\frac{1}{5}$, a die can be rolled. The numbers 1 through 5 would represent the outcomes. If a 6 is rolled, it is ignored.

For experiments with more than six outcomes, other devices can be used. For example, there are dice for games that have five sides, eight sides, 10 sides, etc. (Again, the best device to use is a random number generator such as a computer or calculator or even a table of random numbers.)

EXAMPLE

Simulate the genders of a family with five children.

SOLUTION

Five coins can be tossed. A head represents a male and a tail represents a female. For example, the outcome HTHHT represents three boys and two girls. Perform the experiment 10 times to represent the genders of the children of 10 families.

NOTE   *The probability of a male or a female birth is not exactly $\frac{1}{2}$; however, it is close enough for this situation.*

The results are shown in the table.

Trial	Outcome	Number of Boys
1	TTHTT	1
2	TTTTT	0
3	HHTTH	3
4	THTTH	2
5	TTHTH	2
6	HHHHH	5
7	HTHHT	3
8	THHHH	4
9	THHTH	3
10	THTTH	2

Results:

No. of boys	0	1	2	3	4	5
No. of families	1	1	3	3	1	1

In this case, there was one family with no boys and one family with one boy. Three families had two boys. Three families had three boys. One family had four boys, and one family had five boys. The average is 2.5 boys per family of five. Simulation techniques are not perfect, but most of the time they provide a good estimate.

More complicated problems can be simulated as shown next.

EXAMPLE

Suppose a prize is given under a bottle cap of a soda; however, only one in five bottle caps has the prize. Find the average number of bottles that would have to be purchased to win the prize. Use 20 trials.

✔SOLUTION

A die can be rolled until a certain arbitrary number, say 3, appears. The 3 would be the winner. The probability of getting a winner is $\frac{1}{5}$. The number of rolls is tallied. The experiment can be done 20 times. (In general, the more times the experiment is performed, the better the approximation will be.) In this case, if a 6 is rolled, it is not counted. The results are shown next.

Trial	Number of Rolls Until a 3 Was Obtained
1	1
2	6
3	5
4	4
5	11
6	5
7	1
8	3
9	7
10	2
11	4
12	2
13	4
14	1
15	6
16	9
17	1
18	5
19	7
20	11

Now, the average of the number of rolls is 4.75. The theoretical average is 5.

EXAMPLE

A box contains two $1 bills, three $5 bills, and one $10 bill. A person selects a bill at random. Find the expected value of the bill. Perform the experiment 20 times.

SOLUTION

A die can be rolled. If a 1 or a 2 comes up, assume the person wins $1. If a 3, 4, or 5 comes up, assume the person wins $5. If a 6 comes up, assume the person wins $10.

Trial	Number	Amount
1	3	$5
2	6	$10
3	3	$5
4	6	$10
5	4	$5
6	1	$1
7	6	$10
8	4	$5
9	4	$5
10	3	$5
11	6	$10
12	1	$1
13	2	$1
14	5	$5
15	5	$5
16	3	$5
17	1	$1
18	2	$1
19	6	$10
20	3	$5

The average of the amount won is $5.25.

The theoretical average or expected value can be found by using the formula shown in Chapter 5.

$$E(X) = \frac{1}{3}(\$1) + \frac{1}{2}(\$5) + \frac{1}{6}(\$10) = \$4.50.$$

The $5.25 is somewhat better than the theoretical average.

PRACTICE

Explain how a simulation procedure will give you an appropriate answer.

1. In a certain prize give away, you get to scratch a ticket each time you purchase 10 or more gallons of gasoline. In order to win a prize, you must spell the word "WIN." The tickets are printed so that 70% of them have a "W,"

20% have an "I," and 10% have an "N." If a person gets 10 tickets, find the probability that he or she will win the prize.

2. Mary drives to work each day, and there are three traffic lights on the route. The probability that a light is green is $\frac{1}{3}$, and the probability that a light is not green is $\frac{2}{3}$. Assume each light is independent of the other lights. Find the probability that Mary will get all green lights when she drives to work.

3. Two people throw balls at a pin in an amusement park game. Sally hits the pin 70% of the times she throws, and Bill hits the pin 40% of the times he throws. The game is won by the person who hits the target first. If Bill throws first each time, find the probability he will win.

4. Sam needs to wear a tie to work three days a week. He selects a tie at random each morning. Find the probability that he will wear the same tie to work two or more days a week if he owns six different ties.

5. A basketball player has a 50% chance of making a foul shot. If he shoots two foul shots each time, find the probability that he will miss both of them.

 ANSWERS

1. Use a deck of playing cards. Remove the jacks, queens, and kings. Select 10 cards, replacing each card after it has been selected, and record its denomination. Mix the cards after each selection. Let ace through 7 be a "W," 8 and 9 be an "I," and a 10 be the "N." Repeat the experiment 50 times and see how many times you win.

NOTE   *If you get the "WIN" combination before selecting all 10 cards, you can stop and then start over. Divide the number of times you win by the total number of times you perform the experiment.*

2. Roll a die three times. If you get a 1 or 2, consider these two numbers as green lights. Consider 3, 4, 5, or 6 as yellow or red lights. Count the number of times you get all green lights and divide by the total number of times you perform the experiment.

3. Use 10 cards, ace through 10, and for Bill consider ace through 4 as a hit and any other card as a miss. For Sally, consider ace through 7 a hit and 8, 9, and 10 as a miss. Mix the cards and select a card for Bill. If it's a hit, he wins. If it's a miss, then replace the card, mix the cards, and select a card for Sally. If she wins, the game is over. If she misses,

replace the card, mix them up, and select a card for Bill. Continue until someone wins. Record the number of times Bill wins and divide by the total number of times the experiment has been done.

4. Roll a die three times for the events of any given week. For every instance that you obtain the same digit two or three times out of the three rolls, consider this result as "wearing the same tie." After repeating the experiment n times, divide the number of same-digit instances by n.

5. Toss two coins. Consider a head as a shot made and a tail as a shot missed. Count the number of times you got two tails and divide by the total number of times you have performed the experiment.

PROBABILITY SIDELIGHT: Probability in Our Daily Lives

People engage in all sorts of gambles, not just betting money at a casino or purchasing a lottery ticket. People also bet their lives by engaging in unhealthy activities such as smoking, drinking, using drugs, and exceeding the speed limit when driving. A lot of people don't seem to care about the risks involved in these activities, or they don't understand the concepts of probability.

Statisticians (called actuaries) who work for insurance companies can calculate the probabilities of dying from certain causes. For example, based on the population of the United States, the risks of dying from various causes are shown here.

Motor vehicle accident	1 in 7000
Shot with a gun	1 in 10,000
Crossing a street	1 in 60,000
Struck by lightning	1 in 3,000,000
Shark attack	1 in 300,000,000

The risks of dying from various diseases are shown here.

Heart attack	1 in 400
Cancer	1 in 600
Stroke	1 in 2000

As you can see, the probability of dying from a disease is much higher than the probability of dying from an accident.

Another thing that people tend to do is fear situations or events that have a relatively small chance of happening and overlook situations or events that have a higher chance of happening. For example, James Walsh, in his book entitled *How Risk Affects Your Everyday Life*, states that if a person is 20% overweight, the loss of life expectancy is 900 days (about 3 years), whereas the loss of life expectancy from exposure to radiation emitted by nuclear power plants is 0.02 days. So you can see it is much more unhealthy being 20% overweight than it is living close to a nuclear power plant. One of the reasons for this phenomenon is that the media tend to sensationalize certain news events such as floods, hurricanes, and tornadoes, and downplay other less newsworthy events such as smoking, drinking, and being overweight.

In summary, then, when you make a decision or plan a course of action based on probability, get the facts from a reliable source, weigh the consequences of each choice of action, and then make your decision. Be sure to consider as many alternatives as you can.

Summary

Random numbers can be used to simulate many real-life situations. The basic method of simulation is the Monte Carlo method. The purpose of simulation is to duplicate situations that are too dangerous, too costly, or too time consuming to study in real life. Most simulation techniques are done on a computer. Computers enable the person to generate random numbers, tally the results, and perform any necessary computation.

QUIZ

1. Two people who developed simulation techniques are
   A. Fermat and Pascal
   B. Laplace and DeMoivre
   C. Von Neuman and Ulam
   D. Plato and Aristotle

2. Mathematical simulation techniques use _____ numbers.
   A. Prime
   b. Odd
   C. Even
   D. Random

3. The simulation techniques explained in this chapter use the _____ method.
   A. Monte Carlo
   B. Casino
   C. Coin/Die
   D. Tally

4. A coin can be used as a simulation device when there are two outcomes and each outcome has a probability of
   A. $\frac{1}{4}$
   B. $\frac{1}{2}$
   C. $\frac{1}{3}$
   D. $\frac{1}{6}$

5. Which device will **not** generate random numbers?
   A. Computer
   B. Abacus
   C. Dice
   D. Calculator

6. If a die is used to simulate random numbers, each face can be used as a probability of
   A. $\frac{1}{2}$
   B. $\frac{1}{4}$
   C. $\frac{1}{3}$
   D. $\frac{1}{6}$

7. Using a deck of cards for a simulation, each suit would have a probability of
   A. 10%
   B. 20%
   C. 25%
   D. 50%

8. Simulations can be used when the real-life situation is
   A. Dangerous
   B. Expensive
   C. Time consuming
   D. All of the above

9. In order to get more accurate probabilities, the experiment should be done
   A. Once
   B. Twice
   C. A few times
   D. Many times

10. Simulation experiments can be done using
    A. Coins
    B. Dice
    C. Random numbers
    D. All of the above

Game Theory

Probability is used in what is called *game theory*. **Game theory** was developed by John Von Neumann and is a mathematical analysis of games. In many cases, game theory uses probability. In a broad sense, game theory can be applied to sports such as football and baseball, video games, board games, gambling games, investment situations, and even warfare.

CHAPTER OBJECTIVES

In this chapter, you will learn

- The basic concepts of game theory
- How to determine a probability strategy for a two-person game

Two-Person Games

A simplified definition of a **game** is that it is a contest between two players that consists of rules on how to play and how to determine the winner. A game also consists of a *payoff*. A **payoff** is a reward for winning the game. In many cases it is money, but it could be points or even just the satisfaction of winning.

Most games consist of *strategies*. A **strategy** is a rule that determines a player's move or moves in order to win the game or maximize the player's payoff. When a game consists of the loser paying the winner, it is called a **zero sum game**. This means that the sum of the payoffs is zero. For example, if a person loses a game and that person pays the winner $5, the loser's payoff is –$5 and the winner's payoff is +$5. Hence, the sum of the payoff is –$5 + $5 = $0.

Consider a simple game in which there are only two players and each player can make only a finite number of moves. Both players make a move simultaneously and the outcome or payoff is determined by the pair of moves. An example of such a game is called, "rock-paper-scissors." Here each player places one hand behind his or her back, and at a given signal, brings his or her hand out with either a fist, symbolizing "rock," two fingers out, symbolizing "scissors," or all five fingers out symbolizing "paper." In this game, scissors cut paper, so scissors win. A rock breaks scissors, so the rock wins, and paper covers rock, so paper wins. Rock-rock, scissors-scissors, and paper-paper are ties and neither person wins. Now suppose there are two players, say Player A and Player B, and they decide to play for $1. The game can be symbolized by a rectangular array of numbers called a **payoff table**, where the rows represent Player A's moves and the columns represent Player B's moves. If Player A wins, he gets $1 from Player B. If Player B wins, Player A pays him $1, represented by –$1. The payoff table for the game is

Player A's moves	Player B's moves		
	Rock	Paper	Scissors
Rock	0	–$1	$1
Paper	$1	0	–$1
Scissors	–$1	$1	0

For example, if Player A plays paper and Player B plays scissors, Player A loses $1, denoted by –$1. Player B wins because scissors cut paper.

This game can also be represented by a tree diagram, as shown in Figure 11-1.

Player A's move	Player B's move	Payoff
Rock	Rock	0
	Paper	−$1
	Scissors	$1
Paper	Rock	$1
	Paper	0
	Scissors	−$1
Scissors	Rock	−$1
	Paper	$1
	Scissors	0

FIGURE 11-1

Now consider a second game. Each player has two cards. One card is black on one side, and the other card is white on one side. The backs of all four cards are the same, so when a card is placed face down on a table, the color on the opposite side cannot be seen until it is turned over. Both players select a card and place it on the table face down; then they turn the cards over. If the result is two black cards, Player A wins $5. If the result is two white cards, Player A wins $1. If the results are one black card and one white card, Player B wins $2. A payoff table for the game would look like this:

	Player B's card	
Player A's card	Black	White
Black	$5	−$2
White	−$2	$1

The tree diagram for the game is shown in Figure 11-2.

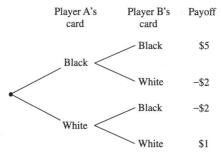

Player A's card	Player B's card	Payoff
Black	Black	$5
	White	−$2
White	Black	−$2
	White	$1

FIGURE 11-2

Player A thinks, "What about a strategy? I will play my black card and hope Player B plays her black card, and I will win $5. But maybe Player B knows this and she will play her white card, and I will lose $2. So, I better play my white card and hope Player B plays her white card, and I will win $1. But she might realize this and play her black card! What should I do?"

In this case, Player A decides that he should play his black card some of the time and his white card some of the time. But how often should he play his black card?

This is where probability theory can be used to solve Player A's dilemma. Let p = the probability that Player A plays his black card on each turn; then $1 - p$ = the probability that Player A plays his white card on each turn. If Player B plays her black card, Player A's expected profit is $\$5 \cdot p - \$2(1 - p)$. If Player B plays her white card, Player A's expected profit is $-\$2p + \$1(1 - p)$, as shown in the table.

	Player B's card	
Player A's card	Black	White
Black	$5p$	$-\$2p$
White	$-\$2(1 - p)$	$\$1(1 - p)$
	$\$5p - \$2(1 - p)$	$-\$2p + \$1(1 - p)$

Now in order to plan a strategy so that Player B cannot outthink Player A, the two expressions should be equal. Hence,

$$5p - 2(1 - p) = -2p + 1(1 - p)$$

Using algebra, we can solve for p:

$$5p - 2(1 - p) = -2p + 1(1 - p)$$
$$5p - 2 + 2p = -2p + 1 - p$$
$$7p - 2 = -3p + 1$$
$$7p + 3p - 2 = -3p + 3p + 1$$
$$10p - 2 = 1$$
$$10p - 2 + 2 = 1 + 2$$
$$10p = 3$$
$$\frac{10p}{10} = \frac{3}{10}$$
$$p = \frac{3}{10}$$

Hence, Player A should play his black card $\frac{3}{10}$ of the time and his white card $\frac{7}{10}$ of the time. His expected gain, no matter what Player B does, when $p = \frac{3}{10}$ is

$$5p - 2(1 - p) = 5 \cdot \frac{3}{10} - 2\left(1 - \frac{3}{10}\right)$$

$$= \frac{15}{10} - 2\left(\frac{7}{10}\right)$$

$$= \frac{15}{10} - \frac{14}{10}$$

$$= \frac{1}{10} \text{ or } \$0.10$$

The same value obtained by substituting $p = \frac{3}{10}$ into $-2p + 1(1 - p)$.

On average, Player A will win \$0.10 per game no matter what Player B does.

Still Struggling

You may wonder why the equation $5p - 2(1 - p) = -2p + 1(1 - p)$ used previously gives an appropriate value for p. Here's why:

Player A wants his expected profit to exceed 0, if possible, in either case; hence, $5p - 2(1 - p) > 0$ and $-2p + 1(1 - p) > 0$. Solving each inequality for p, you get $p > \frac{2}{7}$ and $p < \frac{1}{3}$. Thus, the value of p should be $\frac{2}{7} < p < \frac{1}{3}$.

Now when the equation $5p - 2(1 - p) = -2p + 1(1 - p)$ is solved for p, you get $p = \frac{3}{10}$ and $\frac{2}{7} < \frac{3}{10} < \frac{1}{3}$ since $\frac{60}{210} < \frac{63}{210} < \frac{70}{210}$. Hence, the value for p, i.e., $\frac{3}{10}$, satisfies the inequality.

Now Player B decides she better figure her expected loss no matter what Player A does. Using similar reasoning, the table will look like this when the probability that Player B plays her black card is s, and her white card with probability $1 - s$.

Player A's card	Player B's card		
	Black	White	
Black	$5s$	$-\$2(1-s)$	$\$5s - \$2(1-s)$
White	$-\$2s$	$\$1(1-s)$	$-\$2(s) + \$1(1-s)$

Solving for s when both expressions are equal, we get:

$$5s - 2(1-s) = -2(s) + 1(1-s)$$
$$5s - 2 + 2s = -2s + 1 - s$$
$$7s - 2 = -3s + 1$$
$$7s + 3s - 2 = -3s + 3s + 1$$
$$10s - 2 = 1$$
$$10s - 2 + 2 = 1 + 2$$
$$10s = 3$$
$$\frac{10s}{10} = \frac{3}{10}$$
$$s = \frac{3}{10}$$

So Player B should play her black card $\frac{3}{10}$ of the time and her white card $\frac{7}{10}$ of the time. Player B's payout when $s = \frac{3}{10}$ is

$$\$5s - \$2(1-s) = 5\left(\frac{3}{10}\right) - \$2\left(\frac{7}{10}\right)$$
$$= \frac{15}{10} - \frac{14}{10}$$
$$= \frac{1}{10} \text{ or } \$0.10$$

The same value obtained if $s = \frac{3}{10}$ is placed into $-2s + 1(1-s)$.

Hence the maximum amount that Player B will lose on average is $0.10 per game no matter what Player A does.

When both players use their strategy, the results can be shown by combining the two tree diagrams and calculating Player A's expected gain as shown in Figure 11-3.

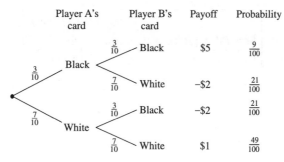

FIGURE 11-3

Hence, Player A's expected gain is

$$\$5\left(\frac{9}{100}\right) - \$2\left(\frac{21}{100}\right) - \$2\left(\frac{21}{100}\right) + \$1\left(\frac{49}{100}\right)$$

$$\frac{45}{100} - \frac{42}{100} - \frac{42}{100} + \frac{49}{100} = \frac{10}{100} = \$0.10$$

The number $0.10 is called the **value** of the game. If the value of the game is 0, then the game is said to be fair.

The *optimal strategy* for Player A is to play the black card $\frac{3}{10}$ of the time and the white card $\frac{7}{10}$ of the time. The optimal strategy for Player B is the same in this case.

The **optimal strategy** for Player A is defined as a strategy that can guarantee him an average payoff of V(the value of the game) no matter what strategy Player B uses. The **optimal strategy** for Player B is defined as a strategy that prevents Player A from obtaining an average payoff greater than V(the value of the game) no matter what strategy Player A uses.

NOTE *When a player selects one strategy some of the time and another strategy at other times, it is called a* **mixed** *strategy, as opposed to using the same strategy all of the time. When the same strategy is used all of the time, it is called a* **pure** *strategy.*

EXAMPLE

Two generals, A and B, decide to play a game. General A can attack General B's city either by land or by sea. General B can defend either by land or sea. They agree on the following payoff for General A.

	General B (defend)	
General A (attack)	Land	Sea
Land	–$50	$150
Sea	$180	–$100

Find the optimal strategy for each player and the value of the game.

SOLUTION

Let p = probability of attacking by land and $1 - p$ = probability of attacking by sea.

If General A attacks by land and General B defends by land, General A loses $50 (i.e., –$50). If General A attacks by land and General B defends by sea, General A wins $150. If General A attacks by sea and General B defends by land, General A wins $180. If General A attacks by sea and General B defends by sea, General A loses $100 (i.e., –$100).

General A's expected payoff if he attacks and General B defends by land is –50p$ + $180 (1 – p), and if General B defends by sea is 150p$ – $100 (1 – p). Equating the two and solving for p, we get

$$-50p + 180(1 - p) = 150p - 100(1 - p)$$
$$-50p + 180 - 180p = 150p - 100 + 100p$$
$$-230p + 180 = 250p - 100$$
$$-230p - 250p + 180 = 250p - 250p - 100$$
$$-480p + 180 = -100$$
$$-480p + 180 - 180 = -100 - 180$$
$$-480p = -280$$
$$\frac{-480p}{-480} = \frac{-280}{-480}$$
$$p = \frac{7}{12}$$

Hence, the optimal strategy for General A is to attack by land $\frac{7}{12}$ of the time and by sea $\frac{5}{12}$ of the time. The value of the game is

$$-50p + 180(1 - p)$$
$$-50\left(\frac{7}{12}\right) + 180\left(\frac{5}{12}\right)$$
$$= \$45.83 \text{ (rounded)}$$

Now let's figure out General B's strategy.

General B should defend by land with a probability of s, and by sea with a probability of $1 - s$. Hence,

$$-50s + 150(1 - s) = 180s - 100(1 - s)$$

$$-50s + 150 - 150s = 180s - 100 + 100s$$

$$-200s + 150 = 280s - 100$$

$$-480s + 150 = -100$$

$$-480s + 150 - 150 = -100 - 150$$

$$-480s = -250$$

$$\frac{-480s}{-480} = \frac{-250}{-480}$$

$$s = \frac{25}{48}$$

Hence, General B should defend by land with a probability of $\frac{25}{48}$ and by sea with a probability of $\frac{23}{48}$.

In other words, General B's optimal strategy is to defend by land 25 out of 48 times, and by sea 23 out of said 48 times.

A tree diagram for this problem is shown in Figure 11-4.

A payoff table can also consist of probabilities. This type of problem is shown in the next example.

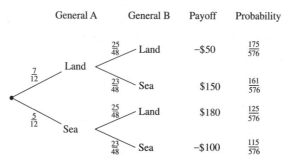

FIGURE 11-4

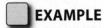EXAMPLE

Player A and Player B decided to play one-on-one basketball. Player A can take either a long shot or a lay-up shot. Player B can defend against either one. The payoff table shows the probabilities of a successful shot for each situation. Find the optimal strategy for each player and the value of the game.

	Player B (defense)	
Player A (offense)	Long shot	Lay-up shot
Long shot	0.2	0.5
Lay-up shot	0.6	0.1

SOLUTION

Let p be the probability of shooting a long shot and $1 - p$ the probability of shooting a lay-up shot. Then the probability of making a shot against a long shot defense is $0.2p + 0.6(1 - p)$ and against a lay-up defense is $0.5p + 0.1(1 - p)$. Equating and solving for p, we get

$$0.2p + 0.6(1 - p) = 0.5p + 0.1(1 - p)$$

$$0.2p + 0.6 - 0.6p = 0.5p + 0.1 - 0.1p$$

$$-0.4p + 0.6 = 0.4p + 0.1$$

$$-0.4p - 0.4p + 0.6 = 0.4p - 0.4p + 0.1$$

$$-0.8p + 0.6 = 0.1$$

$$-0.8p + 0.6 - 0.6 = 0.1 - 0.6$$

$$-0.8p = -0.5$$

$$\frac{-0.8p}{-0.8} = \frac{-0.5}{0.8}$$

$$p = \frac{5}{8}$$

Then the probability of making a successful shot when $p = \frac{5}{8}$ is

$$0.2p + 0.6(1 - p) = 0.2\left(\frac{5}{8}\right) + 0.6\left(1 - \frac{5}{8}\right)$$

$$= 0.2\left(\frac{5}{8}\right) + 0.6\left(\frac{3}{8}\right)$$

$$= \frac{1}{8} + \frac{1.8}{8}$$

$$= \frac{10}{80} + \frac{18}{80}$$

$$= \frac{28}{80} = \frac{7}{20}$$

Hence, Player A will be successful $\frac{7}{20}$ of the time. The value of the game is $\frac{7}{20}$. Player B should defend against a long shot $\frac{1}{2}$ of the time and against a lay-up shot $\frac{1}{2}$ of the time. The solution is left as a practice question. See Question 5.

 PRACTICE

1. A simplified version of football can be thought of as two types of plays. The offense can either run or pass and the defense can defend against a running play or a passing play. The playoff yards gained for each play are shown in the payoff box. Find the optimal strategy for each and determine the value of the game.

Offense	Defense	
	Against run	Against pass
Run	2	6
Pass	8	−10

2. For purposes of military training, a player can either hide behind a rock or in a tree. The other player can either select a pistol or a rifle. The probabilities for success are given in the payoff box. Determine the optimal strategy and the value of the game.

Player A	Player B	
	Rock	Tree
Pistol	0.6	0.2
Rifle	0.4	0.7

3. Person A has two cards, an ace (1) and a 3. Person B has two cards, a 2 and a 4. Each person plays one card. If the sum of the cards is 3 or 7, Person B pays Person A $3 or $7, respectively, but if the sum of the cards is 5, Person A pays Person B $5. Construct a payoff table, determine the optimal strategy for each player, and find the value of the game. Is the game fair?

4. A street vendor without a license has a choice to open on Main Street or Railroad Avenue. The city inspector can only visit one location per day. If he catches the vendor, the vendor must pay a $50 fine; otherwise, the vendor can make $100 at Main Street or $75 at Railroad Avenue. Construct the payoff table, determine the optimal strategy for both locations, and find the value of the game.

5. Find the optimal strategy for Player B in the last example (basketball).

ANSWERS

1. Let p be the probability of running and $1 - p$ be the probability of passing. Then

$$2p + 8(1 - p) = 6p - 10(1 - p)$$
$$2p + 8 - 8p = 6p - 10 + 10p$$
$$-6p + 8 = 16p - 10$$
$$-6p - 16p + 8 = 16p - 16p - 10$$
$$-22p + 8 = -10$$
$$-22p + 8 - 8 = -10 - 8$$
$$-22p = -18$$
$$\frac{-22p}{-22} = \frac{-18}{-22}$$
$$p = \frac{9}{11}$$

Hence, the player (offensive) should run $\frac{9}{11}$ of the time and pass $\frac{2}{11}$ of the time. The value of the game is

$$2\left(\frac{9}{11}\right) + 8\left(\frac{2}{11}\right) = \frac{34}{11} = 3\frac{1}{11}$$

Let s = the probability of defending against the run and $1 - s$ = the probability of defending against the pass; then

$$2s + 6(1 - s) = 8s - 10(1 - s)$$
$$2s + 6 - 6s = 8s - 10 + 10s$$
$$-4s + 6 = 8s - 10 + 10s$$
$$-4s + 6 = 18s - 10$$
$$16 = 22s$$
$$8 = 11s$$
$$s = \frac{8}{11}$$

Hence, the player (defense) should defend against the run $\frac{8}{11}$ of the time and pass $\frac{3}{11}$ of the time.

2. Let p be the probability of selecting a pistol and $1 - p$ be the probability of selecting a rifle. Then

$$0.6p + 0.4(1 - p) = 0.2p + 0.7(1 - p)$$

$$0.6p + 0.4 - 0.4p = 0.2p + 0.7 - 0.7p$$

$$0.2p + 0.4 = -0.5p + 0.7$$

$$0.2p + 0.5p + 0.4 = -0.5p + 0.5p + 0.7$$

$$0.7p + 0.4 = 0.7$$

$$0.7p + 0.4 - 0.4 = 0.7 - 0.4$$

$$0.7p = 0.3$$

$$\frac{0.7p}{0.7} = \frac{0.3}{0.7}$$

$$p = \frac{3}{7}$$

The value of the game when $p = \frac{3}{7}$ is $0.6p + 0.4(1 - p) = 0.6(\frac{3}{7}) + 0.4(\frac{4}{7}) = \frac{17}{35}$.

When Player A selects a pistol $\frac{3}{7}$ of the time, he will be successful $\frac{17}{35}$ of the time.

Let s = the probability of Player B hiding behind a rock and $1 - s$ = the probability of hiding in a tree; then

$$0.6s + 0.2(1 - s) = 0.4s + 0.7(1 - s)$$

Multiply the above equation by 10, and obtain

$$6s + 2(1 - s) = 4s + 7(1 - s)$$

$$6s + 2 - 2s = 4s = 7 - 7s$$

$$4s + 2 = -3s + 7$$

$$7s = 5$$

$$s = \frac{5}{7} \text{ whence } 1 - s = \frac{7}{7} - \frac{5}{7} = \frac{2}{7}$$

Hence, Player B should hide behind a rock $\frac{5}{7}$ of the time and in a tree $\frac{2}{7}$ of the time.

3. The payoff table is

	Player B	
Player A	Two	Four
Ace	3	−5
Three	−5	7

Let p be the probability that Player A plays the ace and $1 - p$ be the probability that Player A plays the 3. Then

$$3p - 5(1 - p) = -5p + 7(1 - p)$$
$$3p - 5 + 5p = -5p + 7 - 7p$$
$$8p - 5 = -12p + 7$$
$$8p + 12p - 5 = -12p + 12p + 7$$
$$20p - 5 + 5 = 7 + 5$$
$$20p = 12$$
$$p = \frac{12}{20} = \frac{3}{5}$$

The value of the game when $p = \frac{3}{5}$ is

$$3p - 5(1 - p) = 3\left(\frac{3}{5}\right) - 5\left(1 - \frac{3}{5}\right)$$
$$= 3\left(\frac{3}{5}\right) - 5\left(\frac{2}{5}\right)$$
$$= -\frac{1}{5} \text{ or } -\$0.20$$

Player A will lose on average $0.20 per game. Thus, the game is not fair.

Let s be the probability that Player B plays the 2 and $1 - s$ be the probability that Player B plays the 4; then

$$3s - 5(1 - s) = -5s + 7(1 - s)$$
$$3s - 5 + 5s = -5s + 7 - 7s$$
$$8s - 5 = -12s + 7$$
$$8s + 12s - 5 = -12s + 12s + 7$$
$$20s - 5 = 7$$
$$20s - 5 + 5 = 7 + 5$$

$$20s = 12$$

$$\frac{20s}{20} = \frac{12}{20}$$

$$s = \frac{12}{20} = \frac{3}{5} \text{ and } 1 - s = 1 - \frac{3}{5} = \frac{2}{5}.$$

Player B should play the two, 3 times out of 5; and should play the four 2 times out of 5.

4. The payoff table is

	Inspector	
Vendor	Main St.	Railroad Ave.
Main St.	−$50	$100
Railroad Ave.	$75	−$50

Let p be the probability that the vendor selects Main St. and $1 - p$ be the probability that the vendor selects Railroad Ave. Then,

$$-50p + 75(1 - p) = 100p - 50 (1 - p)$$

$$-50p + 75 - 75p = 100p - 50 + 50p$$

$$-125p + 75 = 150p - 50$$

$$-125p - 150p + 75 = 150p - 150p - 50$$

$$-275p + 75 = -50$$

$$-275p + 75 - 75 = -50 - 75$$

$$-275p = -125$$

$$\frac{-275p}{-275} = \frac{-125}{-275}$$

$$p = \frac{125}{275} = \frac{5}{11}$$

The value of the game when $p = \frac{5}{11}$ is

$$-50p + 75(1 - p) = -50\left(\frac{5}{11}\right) + 75\left(1 - \frac{5}{11}\right)$$

$$= -50\left(\frac{5}{11}\right) + 75\left(\frac{6}{11}\right)$$

$$= 18\frac{2}{11} = 18.18 \text{ (rounded)}$$

Thus, if the vendor selects Main St. 5 times out of 11, he will make $18.18.

Let s be the probability that the inspector shows up at Main St. and $1 - s$ be the probability that the inspector shows up at Railroad Avenue. Then

$$-50s + 100(1 - s) = 75s - 50(1 - s)$$

$$-50s + 100 - 100s = 75s - 50 + 50s$$

$$-150s + 100 = 125s - 50$$

$$-150s - 125s + 100 = 125s - 125s - 50$$

$$-275s + 100 = -50$$

$$-275s + 100 - 100 = -50 - 100$$

$$-275s = -150$$

$$\frac{-275s}{-275} = \frac{-150}{-275}$$

$$s = \frac{6}{11}$$

The inspector's optimal strategy will be to show up at Main St. 6 times out of 11 and at Railroad Avenue. 5 times out of 11.

5. Let s be the probability that Player B defends against the long shot and $1 - s$ be the probability that Player B defends against the lay-up shot. Then

$$0.2s + 0.5(1 - s) = 0.6s + 0.1(1 - s)$$

$$0.2s + 0.5 - 0.5s = 0.6s + 0.1 - 0.1s$$

$$-0.3s + 0.5 = 0.5s + 0.1$$

$$-0.3s - 0.5s + 0.5 = 0.5s - 0.5s + 0.1$$

$$-0.8s + 0.5 = 0.1$$

$$-0.8s + 0.5 - 0.5 = 0.1 - 0.5$$

$$-0.8s = -0.4$$

$$\frac{-0.8s}{-0.8} = \frac{-0.4}{-0.8}$$

$$s = \frac{1}{2}$$

Hence, Player B must defend against a long shot $\frac{1}{2}$ of the time and against a lay-up shot $\frac{1}{2}$ of the time.

PROBABILITY SIDELIGHT: **Computers and Game Theory**

Computers have been used to analyze games, most notably the game of chess. Experts have written programs enabling computers to play humans. Matches between chess champion Garry Kasparov and the computer named Deep Blue, as well as his matches against the newer computer X3D Fritz, have received universal notoriety.

Computers cannot think, but they can make billions of calculations per second. What the computer does when it is its turn to make a chess move is to generate a tree of moves. Each player has about 20 choices of a move per turn. Based on these choices, the computer calculates the possible moves of its human opponent; then it makes a move based on the human's possible moves.

With each move, the computer evaluates the position of the chess pieces on the board at that time. Each chess piece is assigned a value based on its importance. For example, a pawn is worth 1 point, a knight is worth 3 points, a rook is worth 5 points, and a queen is worth 9 points. The computer then works backward, assuming its human opponent will make his best move.

This process is repeated after each human move. It is not possible for the computer to make trees for an entire game since it has been estimated that there are 10^{1050} possible chess moves. By looking ahead several moves, the computer can play a fairly decent game. Some programs can beat almost all human opponents. (Chess champions excluded, of course!)

As the power of the computer increases, the more trees the computer will be able to evaluate within a specific time period. Also, computers have been able to be programmed to remember previous games, thus helping in its analysis of the trees. Many people think that in the future, a computer will be built that will be able to defeat all human players.

Summary

Game theory uses mathematics to analyze games. These games can range from simple board games to warfare. A game can be considered a contest between two players that consists of rules on how to play the game and how to determine the winner. In this chapter, only two-player, zero-sum games were explained. A payoff table is used to determine how much a person wins or loses. Payoff tables can also consist of probabilities.

A strategy is a rule that determines a player's move or moves in order to win the game or maximize the player's payoff. An optimal strategy is the strategy that a player uses that will guarantee him or her an average payoff of a certain amount no matter what the other player does. An optimal strategy for a player could also be one that will prevent the other player from obtaining an average payoff greater than a certain amount. This amount is called the value of the game. If the value of the game is 0, then the game is fair.

QUIZ

1. **In a game where one player pays the other player and vice versa, the game is called _____ game.**
   A. A zero sum
   B. No win
   C. An even sum
   D. A payoff

2. **The person who developed the concepts of game theory was**
   A. Blasé Pascal
   B. John Von Neumann
   C. Leonhard Euler
   D. Garry Kasparov

3. **The reward for winning the game is called the**
   A. Loss or win
   B. Strategy
   C. Payoff
   D. Bet

4. **If a game is fair, the value of the game will be**
   A. Undetermined
   B. −1
   C. 1
   D. 0

5. **When both players use an optimal strategy, the amount that on average is the payoff over the long run is called the _____ of the game.**
   A. Odds
   B. Strength
   C. Winnings
   D. Value

Use the following payoff table to answer Questions 6 to 10.

	Player B	
Player A	Y	X
X	3	6
Y	5	−9

6. **If Player A chooses Y and Player B chooses X, the payoff is**

   A. 3
   B. 5
   C. 6
   D. −9

7. **The optimal strategy for Player A would be to select X with a probability of**

   A. $\frac{3}{17}$
   B. $\frac{5}{11}$
   C. $\frac{14}{17}$
   D. $\frac{6}{11}$

8. **When Player A plays X using his optimal strategy, the value of the game is**

   A. $4\frac{7}{11}$
   B. $3\frac{6}{17}$
   C. $5\frac{4}{17}$
   D. $2\frac{4}{11}$

9. **The optimal strategy for Player B would be to select Y with a probability of**

   A. $\frac{15}{17}$
   B. $\frac{9}{11}$
   C. $\frac{5}{11}$
   D. $\frac{2}{17}$

10. **When Player B uses his optimal strategy, the value of the game will be**

    A. $4\frac{5}{17}$
    B. $2\frac{8}{11}$
    C. $1\frac{5}{11}$
    D. $3\frac{6}{17}$

chapter **12**

Actuarial Science

An **actuary** is a person who uses mathematics to analyze risks in order to determine insurance rates, investment strategies, and other situations involving future payouts. Most actuaries work for insurance companies; however, some work for the United States government in the Social Security and Medicare programs and others as consultants to business and financial institutions. The main function of an actuary is to determine premiums for life and health insurance policies and retirement accounts, as wells as premiums for flood insurance, mine subsidence, etc. Actuarial science involves several areas of mathematics, including calculus. However, much of actuarial science is based on probability.

CHAPTER OBJECTIVES

In this chapter, you will learn

- How to determine the probability of events related to life insurance using a period life table

- How to determine the payout of the life insurance company that insures a group of people at a certain age

Mortality Tables

Insurance companies collect data on various risk situations, such as life expectancy, automobile accidents, hurricane damages, etc. The information can be summarized in table form. One such table is called a *mortality table* or *a period life table*. You can find one near the end of this chapter. The **mortality table** used here is from the Social Security Administration and shows the ages for males and females, the probability of dying at a specific age, the number of males and females surviving out of 100,000 during a specific year of their lives, and life expectancies for a given age. By "life expectancy" here is meant the remaining number of years in one's life, not the number of years from birth to death (as usually understood). The following examples show how to use the mortality table.

EXAMPLE

Find the probability of a male dying during his 34th year.

SOLUTION

Based on the mortality table, there are 96,500 males out of 100,000 alive at the beginning of year 34 and 96,345 males living at the beginning of year 35, so to find the number of males who have died during year 34, subtract 96,500 – 96,345 = 155. Therefore, 155 out of 96,500 people have died. Next find the probability.

$$P(\text{dying at age 34}) = \frac{\text{number who died during the year}}{\text{number who were alive at the beginning of 34th year}}$$

$$= \frac{155}{96,500} \approx 0.001606$$

(Notice that under the column labeled "Death probability," the figure given for 34-year-old males is 0.001603. The discrepancy is probably because computations for this column were based on sample sizes larger than 100,000, or perhaps it was due to rounding.)

EXAMPLE

On average, how long can one expect a male who is 34 years old to live?

SOLUTION

Looking at the table for 34-year-old males, the last column shows a life expectancy of 43.16 years. This means that at age 34, a male can expect to live on average another 43.16 years or to age (34 + 43.16) = 77.16 years. Interpreting this means that the average of the remaining life expectancies of males age 34 is 43.16 years. Remember, this is an **average**, not a guarantee.

 **EXAMPLE**

Find the death rate for 40-year-old females.

 **SOLUTION**

From the table for a 40-year-old woman, there are 97,564 out of 100,000 living, and for age 41, there are 97,425 females living; hence, 97,564 − 97,425 = 139 females died during their 40th year of life. Now the death rate is 139 females out of a total of 97,562 or

$$P(\text{dying at } 40) = \frac{\text{number who died during the year}}{\text{number living at the beginning of year 40}}$$

$$= \frac{139}{97,564} \approx 0.001425$$

Notice that the table gives a value of 0.001430 under the column "Death probability." The reason for this discrepancy is probably because samples larger than 100,000 females were used in the calculation.

 EXAMPLE

What is the probability that a female age 30 will die before age 60?

 **SOLUTION**

The number of females living at age 30 is 98,454 out of 100,000, and the number of females living at age 60 is 91,109. So to find the number of females who died, subtract the two numbers: 98,454 − 91,109 = 7345.

That is, 7345 females died between age 30 and age 60. Next, find the probability.

$$P = \frac{\text{number who died}}{\text{number living at the beginning of year 30}}$$

$$= \frac{7345}{98,454} \approx 0.0746$$

In other words, there is about a 7.46% chance that a female age 30 will die before age 60.

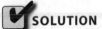 **EXAMPLE**

What is the probability that a male who is 40 will live to the age of 66?

SOLUTION

At age 40, there are 95,431 males out of 100,000 alive. At age 65, there are 77,992 males alive. Hence,

$$P(\text{live to 66}) = \frac{\text{number living at 66}}{\text{number living at 40}} = \frac{77,992}{95,431} \approx 0.817$$

In other words, the probability of a 40-year-old male living to age 66 is 0.817, or about 82%.

EXAMPLE

How many females age 21 will die before age 65?

SOLUTION

At age 21, there are 98,934 females out of 100,000 alive. At age 65, there are 87,217 females alive. Therefore, 98,934 − 87,217 = 11,717 females have died between the ages of 21 and 65. This is out of 98,934 who made it to age 21.

PRACTICE

1. Find the probability that a female will die at age 54.

2. On average, how many more years can a male who is age 32 expect to live?

3. Find the death rate for a 67-year-old female.

4. What is the probability that a male age 48 will live to age 67?

5. What is the probability that a 32-year-old female will live to age 62?

6. Find the probability that a male will live to 18 years of age.

7. How many years longer can a female age 25 expect to live than a male age 25?

8. About how many 2-year-old females will die before they reach 12 years of age?

9. What is the probability that a male age 16 will live to age 55?

10. Find the probability that a male will live to age 68.

ANSWERS

1. From the table, we see the probability that a female age 54 will die is 0.004338.

   Alternate solution: There are 94,067 females out of 100,000 females alive at age 54 and 93,659 females alive at age 55. So, 94,067 − 93,659 = 408 females have died at age 54. So, P(number who died at age 54) = $\frac{\text{number who died at age 54}}{\text{number alive at age 54}} = \frac{408}{94,067} \approx 0.004337$.

2. From the table, a 32-year-old male can expect to live 45.02 years.

3. From the table, the death rate for a 67 year-old-female is 0.013117.

   Alternate solution: At age 67, there are 85,223 out of 100,000 females alive and at age 68, there are 84,105 females alive. So, 85,223 − 84,105 = 1118 have died during the year, so P(dying at age 67) = $\frac{\text{number who died at age 67}}{\text{number who were alive at age 67}} = \frac{1118}{85,223} \approx 0.013119$.

4. At age 48, there are 92,958 out of 100,000 males alive. At age 67, there are 76,540 males alive, so P(living to age 67) = $\frac{\text{number alive at age 67}}{\text{number alive at age 48}} = \frac{76,540}{92,958} \approx 0.823$ or 82.3%.

5. At age 32, there are 98,325 out of 100,000 females alive. At age 62, there are 89,732 out of 100,000 females alive. So, P(living to age 62) = $\frac{\text{number living at age 62}}{\text{number living at age 32}} = \frac{89,732}{98,325} = 0.913 \approx 91.3\%$.

6. At age 18, there are 98,745 males out of 100,000 alive, so P(a male will live to age 18) = $\frac{98,745}{100,000} = 0.987 = 98.7\%$.

7. At age 25, a female can expect to live 56.13 more years. At age 25, a male can expect to live 51.53 more years, so 56.13 − 51.53 = 4.6 years. Hence, a female can expect to live 4.6 years longer than a male if both are 25 years old.

8. At age 2, there are 99,351 females out of 100,000 alive. At age 12, there are 99,210 females alive, so 99,351 – 99,210 = 141 females who are 2 years old and will die before age 12.

9. There are 98,912 males out of 100,000 that are alive at age 16, and 89,037 males alive at age 55. So, $P(\text{living to age 55}) = \frac{\text{number alive at 55}}{\text{number alive at 16}} = \frac{89,037}{98,912} \approx 0.900 = 90\%$.

10. There are 74,993 males out of 100,00 alive at age 68, so $P(\text{living to age 68}) = \frac{74,993}{100,000} \approx 0.7499$ or about 75%.

Life Insurance Policies

There are many different types of life insurance policies. A **straight life insurance policy** requires that you make payments for your entire life. Then when you die, your *beneficiary* is paid the face value of the policy. A **beneficiary** is a person designated to receive the money from an insurance policy.

Another type of policy is a **term policy**. Here the insured pays a certain premium for 20 years. If the person dies during the 20-year period, his or her beneficiary receives the value of the policy. If the person lives beyond the 20-year period, he or she receives nothing. This kind of insurance has low premiums, especially for younger people since the probability of their dying is relatively small.

Another type of life insurance policy is called an **endowment** policy. In this case, if a person purchases a 20-year endowment policy and lives past 20 years, the insurance company will pay the face value of the policy to the insured. Naturally, the premiums for this kind of policy are much higher than those for a term policy.

The table shows the approximate yearly or annual premiums for a $100,000, 20-year term policy. These are based on very healthy individuals. Insurance companies adjust the premiums for people with health problems. Again, remember that the procedure is what is important, not the exact numbers.

Age	Male	Female
21	$115	$105
30	$142	$127
40	$150	$130

EXAMPLE

If a 21-year-old healthy male takes a 20-year term life insurance policy for $100,000, how much would he pay in premiums if he lived at least 20 years?

SOLUTION

His premium would be $115 per year, so he would pay $115 × 20 years = $2300.

EXAMPLE

If a healthy 40-year-old male takes a 20-year term life insurance policy for $25,000, how much would he pay if he lives for at least 20 years?

SOLUTION

The premium for a healthy 40-year-old male for a 20-year term policy of $100,000 is $150. So for a $25,000 policy, the premium can be found by making a ratio equal to $\frac{\text{face value of insurance policy}}{\$100,000}$ and multiplying it by the premium:

$$\frac{\$25,000}{\$100,000} \times \$150 = \$37.50 = \text{annual premium.}$$

Then multiply by 20 years:

$$\$37.55 \times 20 = \$750.$$

EXAMPLE

If the life insurance company insures 100 healthy females age 30 for a 20-year, $100,000 term life insurance policy, find the approximate amount the company will have to pay out.

SOLUTION

First use the mortality table to find the probability that a female aged 30 will die before she reaches age 50. At age 30, there are 98,454 females out of 100,000 living. At age 50, there are 95,460 living. So, in 20 years, 98,454 – 95,460 = 2994 have died during the 20-year period. Hence, the probability of dying is

$$P(\text{dying}) = \frac{\text{number who have died}}{\text{number living at age 30}} = \frac{2994}{98,454} \approx 0.030$$

Hence, about 0.03 or 3% of the females have died during the 20-year period. If the company has insured 100 females, then about 3% × 100 = 3 will die in the 20-year period. The company will have to pay out 3 × $100,000 = $300,000 in the 20-year period.

Notice that knowing this information, the insurance company can estimate its costs (overhead) and calculate premiums to determine its profit.

Another statistic that insurance companies use is called the *median future lifetime* of a group of individuals at a given age. The **median future lifetime** for people living at a certain age is the number of years that approximately one-half of those individuals will still be alive.

EXAMPLE

Find the median future lifetime for a male who is 45 years old.

SOLUTION

Using the mortality table, find the number of males living at age 45. It is 94,075 out of 100,000. Then divide this number by 2 to get 94,075 ÷ 2 = 47,037.5. In the table under the Male/Number of lives column, find the value closest to 47,037.5. It is 47,073. Next find the age of the males that correspond to 47,073. It is 80. In other words, at age 80, about one-half of the males are still living. Subtract 80 – 45 = 35. The median future lifetime of a 45-year-old male is 35 years.

PRACTICE

1. If a healthy 30-year-old female takes a 20-year, $100,000 term life insurance policy, how much would she pay in premiums if she lived to age 50?

2. If a healthy male age 30 takes a 20-year, $40,000 term life insurance policy, about how much would he pay in premiums if he lived to age 50?

3. If a life insurance company insures 100 healthy females age 30 for $50,000, 20-year term policies, how much would the company expect to pay out?

4. Find the median future lifetime of a male who is age 42.

5. Find the median future lifetime of a female who is age 35.

ANSWERS

1. $127 × 20 = $2540

2. $\dfrac{$40,000}{$100,000} × $142 × 20 = 1136

3. At age 30, there are 98,454 females out of 100,000 alive. At age 50, there are 95,460 females alive. Therefore, 98,454 − 95,460 = 2994 females will die.

$$P(\text{dying in 20 years}) = \frac{\text{number who will die}}{\text{number alive at 30}} = \frac{2994}{98,454} = 0.03.$$

Out of 100 females, 100 × 0.03 = 3 will die. Hence, 3 × $100,000 = $300,000 will have to be paid out.

4. At age 42, there are 94,955 males out of 100,000 alive; 94,955 ÷ 2 = 47,477.5. At age 80, there are 47,073 males alive. This is the closest number to 47,037.5. So, 80 − 42 = 38 is the median future lifetime.

5. At age 35, there are 98,096 females out of 100,000 alive; 98,096 ÷2 = 49,048 females alive. At age 84, there are 48,895 females alive, which is the closest number to 49,048. Hence, the median future lifetime of a female age 35 is 84 − 35 = 49 years.

PROBABILITY SIDELIGHT: Early History of Mortality Tables

Surveys and censuses have been around for a long time. Early rulers wanted to keep track of the economic wealth and manpower of their subjects. One of the earliest enumeration records appears in the *Bible* in the *Book of Numbers*. Egyptian and Roman rulers were noted for their surveys and censuses.

In the late 1500s and early 1600s, parish clerks of the Church of England in London began keeping records of the births, deaths, marriages, and baptisms of their parishioners. Many of these were published weekly and summarized yearly. They were called the *Bills of Mortality*. Some even included possible causes of death as well, as could be determined at that time. At best, they were "hit and miss" accounts. If a clerk did not publish the information one week, the figures were included in the next week's summary. Also during this time, people began keeping records of deaths due to the various plagues.

Around 1662, an English merchant, John Graunt (1620–1674), began reviewing the *Bills of Mortality* and combining them into tables. He used records from the years 1604 to 1661 and produced tables that he published in a book entitled

National and Political Observations. He noticed that with the exception of plagues or wars, the number of people that died at a certain age was fairly consistent. He then produced a crude mortality table from this information. After reviewing the data, he drew several conclusions. Some were accurate and some were not.

He stated that the number of male births was slightly greater than the number of female births. He also noticed that, in general, women lived longer than men. He stated that physicians treated about twice as many female patients as male patients, and that the physicians were better able to cure the female patients. From this fact, he concluded that either men were more prone to die from their vices or that men didn't go to the doctor as often as women when they were ill!

For his work in this area, he was given a fellowship in the Royal Society of London. He was the first merchant to receive this honor. Until this time, all members were doctors, noblemen, and lawyers.

Two brothers from Holland, Ludwig and Christiaan Huggens (1629–1695), noticed his work. They expanded on Gaunt's work and constructed their own mortality table. This was the first table that used probability theory and included the probabilities of a person dying at a certain age in his or her life; the table also included the probability of surviving to a certain age.

Later, insurance companies began producing and using mortality tables to determine life expectancies and rates for life insurance.

Period Life Table, 2006 (Modified 2010)						
	Male			Female		
Exact Age	Death Probability[1]	Number of Lives[2]	Life Expectancy	Death Probability[1]	Number of Lives[2]	Life Expectancy
0	0.007349	100,000	75.10	0.006055	100,000	80.21
1	0.000465	99,265	74.66	0.000433	99,395	79.70
2	0.000321	99,219	73.69	0.000276	99,351	78.73
3	0.000244	99,187	72.72	0.000184	99,324	77.75
4	0.000194	99,163	71.74	0.000160	99,306	76.77
5	0.000181	99,144	70.75	0.000144	99,290	75.78
6	0.000174	99,126	69.76	0.000133	99,276	74.79
7	0.000163	99,108	68.77	0.000124	99,262	73.80
8	0.000142	99,092	67.79	0.000113	99,250	72.81
9	0.000112	99,078	66.79	0.000102	99,239	71.82
10	0.000085	99,067	65.80	0.000093	99,229	70.82

(Continued)

Period Life Table, 2006 (Modified 2010)						
	Male			**Female**		
Exact Age	**Death Probability[1]**	**Number of Lives[2]**	**Life Expectancy**	**Death Probability[1]**	**Number of Lives[2]**	**Life Expectancy**
11	0.000085	99,059	64.81	0.000094	99,220	69.83
12	0.000135	99,050	63.81	0.000113	99,210	68.84
13	0.000251	99,037	62.82	0.000153	99,199	67.85
14	0.000416	99,012	61.84	0.000210	99,184	66.86
15	0.000595	98,971	60.86	0.000274	99,163	65.87
16	0.000765	98,912	59.90	0.000335	99,136	64.89
17	0.000928	98,836	58.94	0.000385	99,103	63.91
18	0.001077	98,745	58.00	0.000418	99,064	62.93
19	0.001208	98,638	57.06	0.000438	99,023	61.96
20	0.001343	98,519	56.13	0.000457	98,980	60.99
21	0.001470	98,387	55.20	0.000479	98,934	60.01
22	0.001549	98,242	54.28	0.000497	98,887	59.04
23	0.001567	98,090	53.37	0.000511	98,838	58.07
24	0.001540	97,936	52.45	0.000523	98,787	57.10
25	0.001496	97,785	51.53	0.000536	98,736	56.13
26	0.001459	97,639	50.61	0.000550	98,683	55.16
27	0.001432	97,497	49.68	0.000567	98,629	54.19
28	0.001426	97,357	48.75	0.000588	98,573	53.22
29	0.001436	97,218	47.82	0.000612	98,515	52.25
30	0.001454	97,079	46.89	0.000641	98,454	51.28
31	0.001473	96,938	45.96	0.000677	98,391	50.32
32	0.001504	96,795	45.02	0.000720	98,325	49.35
33	0.001546	96,649	44.09	0.000772	98,254	48.39
34	0.001603	96,500	43.16	0.000833	98,178	47.42
35	0.001673	96,345	42.23	0.000903	98,096	46.46
36	0.001761	96,184	41.30	0.000982	98,008	45.50
37	0.001876	96,014	40.37	0.001073	97,911	44.55
38	0.002021	95,834	39.44	0.001179	97,806	43.59
39	0.002193	95,641	38.52	0.001299	97,691	42.65
40	0.002391	95,431	37.61	0.001430	97,564	41.70
41	0.002607	95,203	36.69	0.001570	97,425	40.76
42	0.002842	94,955	35.79	0.001720	97,272	39.82
43	0.003091	94,685	34.89	0.001878	97,104	38.89

(Continued)

Period Life Table, 2006 (Modified 2010)						
	Male			Female		
Exact Age	Death Probability[1]	Number of Lives[2]	Life Expectancy	Death Probability[1]	Number of Lives[2]	Life Expectancy
44	0.003360	94,392	34.00	0.002046	96,922	37.96
45	0.003646	94,075	33.11	0.002229	96,724	37.04
46	0.003960	93,732	32.23	0.002423	96,508	36.12
47	0.004316	93,361	31.35	0.002622	96,274	35.21
48	0.004721	92,958	30.49	0.002826	96,022	34.30
49	0.005166	92,519	29.63	0.003038	95,750	33.39
50	0.005660	92,041	28.78	0.003275	95,460	32.49
51	0.006171	91,520	27.94	0.003535	95,147	31.60
52	0.006653	90,955	27.11	0.003798	94,811	30.71
53	0.007085	90,350	26.29	0.004061	94,450	29.83
54	0.007498	89,710	25.48	0.004338	94,067	28.94
55	0.007936	89,037	24.66	0.004640	93,659	28.07
56	0.008451	88,331	23.86	0.004993	93,224	27.20
57	0.009063	87,584	23.06	0.005419	92,759	26.33
58	0.009797	86,790	22.26	0.005936	92,256	25.47
59	0.010643	85,940	21.48	0.006534	91,708	24.62
60	0.011599	85,026	20.70	0.007219	91,109	23.78
61	0.012624	84,039	19.94	0.007956	90,452	22.95
62	0.013684	82,978	19.19	0.008698	89,732	22.13
63	0.014759	81,843	18.45	0.009424	88,951	21.32
64	0.015890	80,635	17.72	0.010174	88,113	20.52
65	0.017161	79,354	17.00	0.011009	87,217	19.72
66	0.018610	77,992	16.28	0.011986	86,257	18.94
67	0.020216	76,540	15.58	0.013117	85,223	18.16
68	0.021992	74,993	14.89	0.014430	84,105	17.40
69	0.023966	73,344	14.22	0.015924	82,891	16.64
70	0.026212	71,586	13.55	0.017646	81,571	15.90
71	0.028725	69,710	12.91	0.019544	80,132	15.18
72	0.031450	67,707	12.27	0.021523	78,566	14.47
73	0.034385	65,578	11.65	0.023551	76,875	13.78
74	0.037599	63,323	11.05	0.025717	75,064	13.10
75	0.041267	60,942	10.46	0.028247	73,134	12.43
76	0.045411	58,427	9.89	0.031187	71,068	11.78

(Continued)

Period Life Table, 2006 (Modified 2010)						
	Male			**Female**		
Exact Age	Death Probability[1]	Number of Lives[2]	Life Expectancy	Death Probability[1]	Number of Lives[2]	Life Expectancy
77	0.049921	55,774	9.34	0.034405	68,852	11.14
78	0.054797	52,990	8.80	0.037905	66,483	10.52
79	0.060154	50,086	8.29	0.041808	63,963	9.92
80	0.066266	47,073	7.78	0.046337	61,289	9.33
81	0.073175	43,954	7.30	0.051587	58,449	8.76
82	0.080723	40,737	6.84	0.057503	55,433	8.21
83	0.088916	37,449	6.39	0.064135	52,246	7.68
84	0.097922	34,119	5.97	0.071587	48,895	7.17
85	0.107951	30,778	5.56	0.079984	45,395	6.68
86	0.119182	27,456	5.18	0.089431	41,764	6.22
87	0.131736	24,183	4.81	0.100009	38,029	5.78
88	0.145669	20,998	4.46	0.111773	34,226	5.37
89	0.160978	17,939	4.14	0.124745	30,400	4.98
90	0.177636	15,051	3.84	0.138938	26,608	4.62
91	0.195594	12,378	3.56	0.154348	22,911	4.28
92	0.214792	9,957	3.30	0.170963	19,375	3.98
93	0.235163	7,818	3.07	0.188761	16,062	3.69
94	0.256634	5,979	2.86	0.207711	13,030	3.44
95	0.277945	4,445	2.67	0.226885	10,324	3.20
96	0.298731	3,209	2.51	0.245997	7,982	3.00
97	0.318602	2,251	2.36	0.264731	6,018	2.81
98	0.337164	1,534	2.24	0.282754	4,425	2.65
99	0.354023	1,017	2.12	0.299719	3,174	2.49
100	0.371724	657	2.01	0.317702	2,223	2.35
101	0.390310	413	1.90	0.336764	1,516	2.20
102	0.409825	252	1.80	0.356970	1,006	2.07
103	0.430317	148	1.70	0.378389	647	1.94
104	0.451833	85	1.60	0.401092	402	1.82
105	0.474424	46	1.51	0.425157	241	1.70
106	0.498145	24	1.42	0.450667	138	1.59
107	0.523053	12	1.34	0.477707	76	1.48
108	0.549205	6	1.26	0.506369	40	1.38
109	0.576666	3	1.18	0.536751	20	1.28

(Continued)

Period Life Table, 2006 (Modified 2010)						
	Male			Female		
Exact Age	Death Probability[1]	Number of Lives[2]	Life Expectancy	Death Probability[1]	Number of Lives[2]	Life Expectancy
110	0.605499	1	1.11	0.568956	9	1.19
111	0.635774	0	1.03	0.603094	4	1.10
112	0.667563	0	0.97	0.639279	2	1.02
113	0.700941	0	0.90	0.677636	1	0.94
114	0.735988	0	0.84	0.718294	0	0.86
115	0.772787	0	0.78	0.761392	0	0.79
116	0.811426	0	0.72	0.807076	0	0.72
117	0.851998	0	0.66	0.851998	0	0.66
118	0.894598	0	0.61	0.894598	0	0.61
119	0.939328	0	0.56	0.939328	0	0.56

[1]Probability of dying within one year.
[2]Number of survivors out of 100,000 born alive.

Summary

This chapter introduces some of the concepts used in actuarial science. An actuary is a person who uses mathematics in order to determine insurance rates, investment strategies, retirement accounts, and other situations involving future payouts.

Actuaries use mortality tables to determine the probabilities of people living to certain ages. A mortality table shows the number of people out of 1000, 10,000, or 100,000 living at certain ages. It can also show the probability of dying at any given age. Barring unforeseen catastrophic events such as wars, plagues, and such, the number of people dying at a specific age is relatively constant for certain groups of people.

In addition to life insurance, mortality tables are used in other areas. Some of these include social security and retirement accounts.

QUIZ

1. The probability of a male age 37 dying before age 51 is
   A. 0.953
   B. 0.964
   C. 0.036
   D. 0.047

2. The probability that a female age 84 will die is
   A. 0.064
   B. 0.053
   C. 0.037
   D. 0.072

3. The life expectancy of a female who is 53 is
   A. 31.60 years
   B. 30.25 years
   C. 29.83 years
   D. 32.49 years

4. The probability that a male age 24 will live to age 48 is
   A. 0.937
   B. 0.949
   C. 0.921
   D. 0.956

5. The probability that a female age 28 will live until age 79 is
   A. 0.723
   B. 0.649
   C. 0.671
   D. 0.582

6. The probability that a male will live to age 58 is
   A. 0.868
   B. 0.847
   C. 0.890
   D. 0.829

7. How much will a healthy 30-year-old female pay for a $50,000, 20-year term policy if she lives to age 50?
   A. $2540
   B. $1856
   C. $1270
   D. $2480

8. If a life insurance company writes 100 males age 30 a $30,000, 20-year term policy, how much will it pay out in 20 years?

   A. $120,000
   B. $60,000
   C. $90,000
   D. $150,000

9. The median future lifetime of a 49-year-old male is

   A. 31 years
   B. 36 years
   C. 24 years
   D. 28 years

10. The median future lifetime of a 33-year-old female is

   A. 42 years
   B. 51 years
   C. 37 years
   D. 63 years

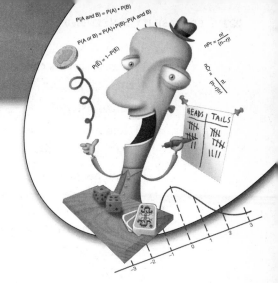

Final Exam

1. If an event cannot occur, its probability is

   A. -1

   B. 1

   C. $\frac{1}{2}$

   D. 0

2. The list of all possible outcomes of a probability experiment is called the

   A. Sample space

   B. Random space

   C. Probability space

   D. Experimental space

3. The probability of an event can be any number from _____ to _____.

   A. 0, infinity

   B. 1, 100

   C. 0, 1

   D. -1, 1

4. When two dice are rolled, the sample consists of _____ outcomes.

   A. 6

   B. 12

   C. 36

   D. 18

5. When two events cannot occur at the same time, they are said to be _____ events.

   A. Inconsistent

   B. Random

   C. Mutually exclusive

   D. Independent

6. If the probability that an event will happen is 0.64, then the probability that the event will not happen is

   A. 0.27

   B. 0.36

   C. 0

   D. −0.64

7. The sum of the probabilities of each outcome in the sample space will always be

   A. 1

   B. $\frac{1}{2}$

   C. 0

   D. Different

8. What is 0!?

   A. 0

   B. 1

   C. Undefined

   D. Infinite

9. When two dice are rolled, the probability of getting a sum of 5 is

   A. $\frac{1}{9}$

   B. 0

   C. $\frac{5}{36}$

   D. $\frac{25}{36}$

10. If 55% of people yawn after they see someone else yawn, find the probability that if three people are randomly selected, they will yawn after seeing someone else yawn.

A. 1.650

B. 0.555

C. 0.330

D. 0.166

11. When a die is rolled, the probability of getting a number less than 3 is
    A. $\frac{1}{12}$
    B. $\frac{1}{3}$
    C. $\frac{2}{3}$
    D. $\frac{1}{6}$

12. When a card is drawn from a deck of 52 cards, the probability of getting a face card is
    A. $\frac{5}{13}$
    B. $\frac{3}{13}$
    C. $\frac{7}{13}$
    D. $\frac{1}{13}$

13. When a die is rolled, the probability of getting an even number less than 5 is
    A. $\frac{1}{6}$
    B. 0
    C. $\frac{1}{3}$
    D. $\frac{1}{2}$

14. A survey conducted at a local restaurant found that 18 people preferred orange juice, 12 people preferred grapefruit juice, and six people preferred apple juice with their breakfasts. If a person is selected at random, the probability that the person will select orange juice is
    A. $\frac{1}{2}$
    B. $\frac{1}{3}$
    C. $\frac{1}{4}$
    D. $\frac{1}{6}$

15. During a sale at a men's store, 14 white shirts, five yellow shirts, six blue shirts, and four green shirts were purchased. If a customer is selected at random, find the probability that the customer purchased a yellow or a blue shirt.

   A. $\frac{7}{30}$

   B. $\frac{19}{30}$

   C. $\frac{9}{30}$

   D. $\frac{11}{30}$

16. A card is selected from an ordinary deck of 52 cards. The probability that it is a 9 or a club is

   A. $\frac{17}{52}$

   B. $\frac{1}{4}$

   C. $\frac{4}{13}$

   D. $\frac{1}{13}$

17. Two dice are rolled; the probability of getting a sum less than or equal to 9 is

   A. $\frac{2}{3}$

   B. $\frac{5}{6}$

   C. $\frac{3}{4}$

   D. $\frac{7}{8}$

18. A card is selected from an ordinary deck of 52 cards. The probability that it is a black 3, 5, or 7 is

   A. $\frac{3}{13}$

   B. $\frac{3}{52}$

   C. $\frac{1}{2}$

   D. $\frac{3}{26}$

19. Three cards are drawn from an ordinary deck of 52 cards without replacement. The probability of getting three aces is

   A. $\frac{1}{169}$

   B. $\frac{1}{5525}$

   C. $\frac{3}{52}$

   D. $\frac{1}{2197}$

20. An automobile license plate consists of three digits followed by three letters. The number of different plates that can be made if repetitions are not permitted is
    A. 11,232,000
    B. 632,450
    C. 17,576,000
    D. 2340

21. The number of different arrangements of the letters of the word *vital* is
    A. 256
    B. 120
    C. 25
    D. 3125

22. A psychology quiz consists of 20 true-false questions. The number of possible ways an answer key can be made is
    A. 40
    B. 400
    C. 380
    D. 1,048,576

23. How many different ways can five cell phones be selected from eight cell phones?
    A. 1680
    B. 40,320
    C. 56
    D. 120

24. The number of different ways four boys and three girls can be selected from six boys and 10 girls is
    A. 1024
    B. 1800
    C. 2400
    D. 560

25. The number of different ways six adults can be seated on a bench is

    A. 6

    B. 36

    C. 720

    D. 46,656

26. A card is selected from an ordinary deck of 52 cards. The probability that it is a red card given that it is a face card is

    A. $\frac{1}{13}$

    B. $\frac{2}{13}$

    C. $\frac{1}{4}$

    D. $\frac{1}{2}$

27. Three cards are selected from an ordinary deck of 52 cards without replacement. The probability of getting all black cards is

    A. $\frac{2}{17}$

    B. $\frac{1}{4}$

    C. $\frac{1}{3}$

    D. $\frac{1}{8}$

28. A die is rolled and a card is drawn from an ordinary deck of 52 cards. The probability of getting a 6 on the die and a spade is

    A. $\frac{1}{10}$

    B. $\frac{3}{4}$

    C. $\frac{4}{15}$

    D. $\frac{1}{24}$

29. When two dice are rolled, the probability of getting a sum of 4 or 9 is

    A. $\frac{7}{36}$

    B. $\frac{13}{36}$

    C. $\frac{1}{3}$

    D. $\frac{1}{6}$

30. The odds in favor of an event when $P(E) = \frac{4}{9}$ are

  A. 4:9

  B. 6:5

  C. 4:5

  D. 5:9

31. The odds against an event when $P(E) = \frac{2}{7}$ are

  A. 2:5

  B. 5:2

  C. 5:7

  D. 7:2

32. The probability of an event when the odds against the event are 7:12 are

  A. $\frac{19}{12}$

  B. $\frac{12}{19}$

  C. $\frac{12}{7}$

  D. $\frac{7}{12}$

33. A person selects a card at random from a box containing five cards. One card has a 10 written on it. Two cards have a 15 written on them, and two cards have a 5 written on them. The expected value of the draw is

  A. 5

  B. 6.5

  C. 8

  D. 10

34. When a game is fair, the odds of winning will be

  A. 3:2

  B. 2:1

  C. 1:1

  D. 1:2

35. A person has six pennies, two nickels, five dimes, and two quarters in her purse. If she selects one coin at random, the expected value of the coin is

    A. 7.7 cents

    B. 6.3 cents

    C. 8.2 cents

    D. 5.4 cents

36. The number of outcomes of a binomial experiment is

    A. 2

    B. 4

    C. 1

    D. Unlimited

37. A survey found that three in 10 Americans say that they have visited a dentist in any given month. If 15 people are selected at random, the probability that exactly five visited a dentist last month is

    A. 0.219

    B. 0.086

    C. 0.155

    D. 0.206

38. A survey found that 60% of shoppers purchased milk when they went grocery shopping. If 12 shoppers are selected at random, the probability that six of them purchased milk is

    A. 0.177

    B. 0.158

    C. 0.053

    D. 0.213

39. A box contains six white balls, two red balls, and four blue balls. A ball is selected at random and its color is noted. It is replaced and another ball is selected. If five balls are selected, the probability that three are white, one is red, and one is blue is

A. $\frac{8}{33}$

B. $\frac{5}{72}$

C. $\frac{5}{36}$

D. $\frac{1}{4}$

40. If there are 300 typographical errors randomly distributed in a 1000-page manuscript, the probability that a given page contains exactly two errors is

A. 0.186

B. 0.033

C. 0.374

D. 0.001

41. A die is rolled; find the probability that an odd number occurs for the first time on the third roll.

A. $\frac{1}{2}$

B. $\frac{1}{8}$

C. $\frac{1}{3}$

D. $\frac{1}{216}$

42. The mean of the standard normal distribution is

A. 1

B. 0

C. 100

D. Variable

43. The percent of the area under the normal distribution curve that falls within 1 standard deviation on either side of the mean is approximately

A. 68

B. 95

C. 99.7

D. Variable

44. The total area under the standard normal distribution curve is

    A. 100%

    B. 95%

    C. 65%

    D. 50%

45. In the graph of the standard normal distribution, the values of the horizontal axis are called

    A. z values

    B. y values

    C. x values

    D. None of the above

46. The scores on a national achievement test are approximately normally distributed with a mean of 120 and a standard deviation of 10. Assume the variable is normally distributed. The probability that a randomly selected student scores between 90 and 140 is

    A. 34.1%

    B. 68.2%

    C. 81.8%

    D. 97.7%

47. The heights of a group of adult males are approximately distributed with a mean of 70 inches and a standard deviation of 2 inches. Assume the variable is normally distributed. The probability that a randomly selected male from the group is between 72 and 76 inches is

    A. 15.9%

    B. 47.7%

    C. 34.1%

    D. 68.2%

48. The average time it takes a subway car to reach its destination is 47 minutes with a standard deviation of 4 minutes. Assume the variable is normally distributed. The probability that it will take the car between 45 minutes and 53 minutes to arrive at its destination is

A. 38.3%

B. 51.6%

C. 62.4%

D. 93.3%

49. The average time it takes a mechanic to fix a flat tire is 12 minutes. The standard deviation is 2 minutes. The variable is approximately normally distributed. Find the probability that a randomly selected tire repair will take less than 7 minutes.

A. 0.006

B. 0.023

C. 0.067

D. 0.184

50. The average monthly food bill for a family is $379. The standard deviation is $20. Assume the variable is normally distributed. If a family is randomly selected, the probability that the food bill is greater than $415 is

A. 3.6%

B. 15.6%

C. 84.4%

D. 94.5%

Use the payoff table for Questions 51 to 55.

	Player B	
Player A	Strategy *X*	Strategy *Y*
Strategy *X*	2	−3
Strategy *Y*	−5	6

51. If Player A uses *X* and Player B uses *Y*, the payoff is

A. 2

B. −3

C. −5

D. 6

52. The optimal strategy for Player A is to use X with a probability of

    A. $\frac{3}{16}$

    B. $\frac{11}{16}$

    C. $\frac{7}{16}$

    D. $\frac{9}{16}$

53. The value of the game is

    A. $\frac{5}{16}$

    B. $-\frac{9}{16}$

    C. $-\frac{3}{16}$

    D. $\frac{13}{16}$

54. The optimal strategy for Player B would be to play Y with a probability of

    A. $\frac{5}{16}$

    B. $\frac{11}{16}$

    C. $\frac{7}{16}$

    D. $\frac{3}{16}$

55. The optimal strategy for Player B would be to play X with a probability of

    A. $\frac{5}{16}$

    B. $\frac{7}{16}$

    C. $\frac{3}{16}$

    D. $\frac{9}{16}$

Use the period life table in Chapter 12 to answer Questions 56 to 60.

56. What is the probability that a female will die at age 50?

    A. 0.008

    B. 0.006

    C. 0.005

    D. 0.003

57. What is the probability that a male age 40 will live to age 60?

    A. 0.807

    B. 0.891

    C. 0.193

    D. 0.099

58. What is the probability that a female age 59 will die before age 68?

    A. 0.083

    B. 0.127

    C. 0.873

    D. 0.917

59. What is the life expectancy of a 33-year-old male?

    A. 51.14 years

    B. 48.39 years

    C. 46.27 years

    D. 44.09 years

60. What is the median future lifetime of a 27-year-old female?

    A. 84 years

    B. 62 years

    C. 57 years

    D. 41 years

Answers to Quizzes and Final Exam

Chapter 1
1. C
2. C
3. A
4. D
5. B
6. D
7. C
8. D
9. D
10. D
11. D
12. B
13. A
14. B
15. C

Chapter 2
1. C
2. B
3. D
4. A
5. C

6. D
7. C
8. A
9. D
10. A
11. B
12. C
13. A
14. A
15. B

Chapter 3
1. C
2. B
3. D
4. C
5. B
6. B
7. B
8. B
9. A
10. B

Chapter 4
1. C
2. B
3. D
4. A
5. D
6. D
7. A
8. C
9. B
10. C

Chapter 5
1. A
2. C
3. D
4. B
5. B
6. A
7. A
8. A
9. C
10. B

11. B
12. B
13. A
14. B
15. C

Chapter 6
1. C
2. B
3. D
4. A
5. C
6. D
7. A
8. C
9. C
10. C
11. D
12. B
13. C
14. D
15. D

Chapter 7
1. C
2. B
3. C
4. A
5. D
6. D
7. C
8. A
9. C
10. D

Chapter 8
1. C
2. D
3. A
4. A
5. C
6. C
7. A
8. B
9. B
10. B
11. D
12. C
13. C
14. B
15. D

Chapter 9
1. B
2. C
3. A
4. A
5. C
6. B

7. B
8. A
9. C
10. B
11. A
12. D
13. C
14. D
15. A

Chapter 10
1. C
2. D
3. A
4. B
5. B
6. D
7. C
8. D
9. D
10. D

Chapter 11
1. A
2. B
3. C
4. D
5. D
6. B
7. C
8. B
9. D
10. D

Chapter 12
1. D
2. D

3. C
4. B
5. B
6. A
7. C
8. D
9. A
10. B

Final Exam
1. D
2. A
3. C
4. C
5. C
6. B
7. A
8. B
9. A
10. D
11. B
12. B
13. C
14. A
15. D
16. C
17. B
18. D
19. B
20. A
21. B
22. D
23. C
24. B
25. C
26. D

27. A
28. D
29. A
30. C
31. B
32. B
33. D
34. C
35. A
36. A
37. D
38. A
39. C
40. B
41. B
42. B
43. A
44. A
45. A
46. D
47. A
48. C
49. A
50. A
51. B
52. B
53. C
54. B
55. D
56. D
57. B
58. A
59. D
60. C

appendix

Bayes' Theorem

A somewhat more difficult topic in probability is called Bayes' theorem.

Given two dependent events, A and B, the previous formulas allowed you to find $P(A \text{ and } B)$ or $P(B|A)$. Related to these formulas is a principle developed by an English Presbyterian minister, Thomas Bayes (1702–1761). It is called *Bayes' theorem*.

Knowing the outcome of a particular situation, Bayes' theorem enables you to find the probability that the outcome occurred as a result of a particular previous event. For example, suppose you have two boxes containing red balls and blue balls. Now if it is known that you selected a blue ball, you can find the probability that it came from Box 1 or Box 2. A simplified version of Bayes' theorem is given next.

For 2 mutually exclusive events, A and B, where event B follows event A,

$$P(A|B) = \frac{P(A) \cdot P(B|A)}{P(A) \cdot P(B|A) + P(\overline{A}) \cdot P(B|\overline{A})}$$

EXAMPLE

Box 1 contains two red balls and one blue ball. Box 2 contains one red ball and three blue balls. A coin is tossed; if it is heads, Box 1 is chosen, and a ball is selected at random. If the toss yields tails, Box 2 is chosen. If the ball is red, find the probability it came from Box 1.

SOLUTION

Let A = selecting Box 1 and $\overline{A}$ = selecting Box 2. Since the selection of a box is based on a coin toss, the probability of selecting Box 1 is $\frac{1}{2}$ and the

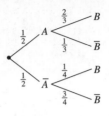

FIGURE A-1

probability of selecting Box 2 is $\frac{1}{2}$; hence, $P(A) = \frac{1}{2}$ and $P(\overline{A}) = \frac{1}{2}$. Let $B =$ selecting a red ball and $\overline{B} =$ selecting a blue ball. From Box 1, the probability of selecting a red ball is $\frac{2}{3}$, and the probability of selecting a blue ball is $\frac{1}{3}$ since there are two red balls and one blue ball. Hence, $P(B|A) = \frac{2}{3}$ and $P(\overline{B}|A) = \frac{1}{3}$. Since there is one red ball in Box 2, $P(B|\overline{A})$ is $\frac{1}{4}$, and since there are 3 blue balls in Box 2, $P(\overline{B}|\overline{A}) = \frac{3}{4}$. The probabilities are shown in Figure A-1.

Hence,
$$P(A|B) = \frac{P(A) \cdot P(B|A)}{P(A) \cdot P(B|A) + P(\overline{A}) \cdot P(B|\overline{A})} = \frac{\dfrac{1}{2} \cdot \dfrac{2}{3}}{\dfrac{1}{2} \cdot \dfrac{2}{3} + \dfrac{1}{2} \cdot \dfrac{1}{4}} = \frac{\dfrac{1}{3}}{\dfrac{1}{3} + \dfrac{1}{8}} = \frac{8}{11}$$

In summary, if a red ball is selected, the probability that it came from Box 1 is $\frac{8}{11}$.

EXAMPLE

Two video products distributors supply video tape boxes to a video production company. Company A sold 100 boxes of which five were defective. Company B sold 300 boxes of which 21 were defective. If a box was defective, find the probability that it came from Company B.

SOLUTION

Let $D =$ the event that "the box is defective" and $\overline{D} =$ the event that "the box is not defective."

Let $P(A) =$ probability that a box selected at random is from Company A; also, let $B = \overline{A}$. Then $P(A) = \frac{100}{400} = \frac{1}{4} = 0.25$; $P(B) = P(\overline{A}) = \frac{300}{400} = \frac{3}{4} = 0.75$. Since there are 5 defective boxes from Company A, $P(D|A) = \frac{5}{100} = 0.05$ and there are 21 defective types from Company B or $\overline{A}$, $P(D|\overline{A}) = \frac{21}{300} = 0.07$. The probabilities are shown in Figure A-2.

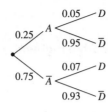

FIGURE A-2

$$P(A|D) = \frac{P(A)\cdot P(D|A)}{P(A)\cdot P(D|A) + P(\bar{A})\cdot P(D|\bar{A})}$$

$$= \frac{(0.25)(0.05)}{(0.25)(0.05) + (0.75)(0.07)}$$

$$= \frac{0.0125}{0.0125 + 0.0525} = \frac{0.0125}{0.065} = 0.192$$

PRACTICE

1. Box 1 contains six green marbles and four yellow marbles. Box 2 contains five yellow marbles and five green marbles. A box is selected at random and a marble is selected from the box. If the marble is green, find the probability that it came from Box 1.

2. An auto parts store purchases rebuilt alternators from two suppliers. From Supplier A, 150 alternators are purchased and 2% are defective. From Supplier B, 250 alternators are purchased and 3% are defective. Given that an alternator is defective, find the probability that it came from Supplier B.

3. Two manufacturers supply paper cups to a catering service. Manufacturer A supplied 100 packages and 5 were damaged. Manufacturer B supplied 50 packages and 3 were damaged. If a package is damaged, find the probability that it came from Manufacturer A.

4. Box 1 contains 10 balls; seven are marked "win" and three are marked "lose." Box 2 contains 10 balls; three are marked "win" and seven are marked "lose." You roll a die. If you get a 1 or 2, you select Box 1 and draw a ball. If you roll 3, 4, 5, or 6, you select Box 2 and draw a ball. Find the probability that Box 2 was selected if you have selected a "win."

5. Using the information in Exercise 4, find the probability that Box 1 was selected if a "lose" was drawn.

ANSWERS

1. $P(B1) = \frac{1}{2}$. Let G = "Marble is green"; then $P(G|B1) = \frac{6}{10} = \frac{3}{5}$. $P(B2) = \frac{1}{2}$;
    $P(G|B2) = \frac{5}{10} = \frac{1}{2}$

$$P(B1|G) = \frac{P(B1) \cdot P(G|B1)}{P(B1) \cdot P(G|B1) + P(B2) \cdot P(G|B2)} = \frac{\frac{1}{2} \cdot \frac{3}{5}}{\frac{1}{2} \cdot \frac{3}{5} + \frac{1}{2} \cdot \frac{1}{2}} = \frac{\frac{3}{10}}{\frac{3}{10} + \frac{1}{4}} = \frac{6}{11}$$

2. $P(A) = \frac{150}{400} = 0.375$; $P(D|A) = 0.02$, where D = "Item is defective"
    $P(B) = \frac{250}{400} = 0.625$
    $P(D|B) = 0.03$

$$P(B|D) = \frac{P(B) \cdot P(D|B)}{P(B) \cdot P(D|B) + P(A) \cdot P(D|A)} = \frac{(0.625)(0.03)}{(0.625)(0.03) + (0.375)(0.02)} = 0.714$$

3. Let D = "Package is damaged." $P(A) = \frac{100}{150} = \frac{2}{3}$; $P(D|A) = \frac{5}{100} = \frac{1}{20}$; $P(B) = \frac{50}{150} = \frac{1}{3}$
    $P(D|B) = \frac{3}{50}$

$$P(A|D) = \frac{P(A) \cdot P(D|A)}{P(A) \cdot P(D|A) + P(B) \cdot P(D|B)} = \frac{\frac{2}{3} \cdot \frac{1}{20}}{\frac{2}{3} \cdot \frac{1}{20} + \frac{1}{3} \cdot \frac{3}{50}} = \frac{\frac{1}{30}}{\frac{1}{30} + \frac{1}{50}} = \frac{5}{8}$$

4. $P(B1) = \frac{1}{3}$; $P(W|B1) = \frac{7}{10}$; $P(B2) = \frac{2}{3}$; $P(W|B2) = \frac{3}{10}$; $P(W|B1) = \frac{7}{10}$

$$P(B2|W) = \frac{P(B2) \cdot P(W|B2)}{P(B2) \cdot P(W|B2) + P(B1) \cdot P(W|B1)} = \frac{\frac{2}{3} \cdot \frac{3}{10}}{\frac{2}{3} \cdot \frac{3}{10} + \frac{1}{3} \cdot \frac{7}{10}} = \frac{\frac{1}{5}}{\frac{1}{5} + \frac{7}{30}} = \frac{6}{13}$$

5. $P(B1) = \frac{1}{3}$; $P(L|B1) = \frac{3}{10}$; $P(B2) = \frac{2}{3}$; $P(L|B2) = \frac{7}{10}$

$$P(B1|L) = \frac{P(B1) \cdot P(L|B1)}{P(B1) \cdot P(L|B1) + P(B2) \cdot P(L|B2)} = \frac{\frac{1}{3} \cdot \frac{3}{10}}{\frac{1}{3} \cdot \frac{3}{10} + \frac{2}{3} \cdot \frac{7}{10}} = \frac{\frac{1}{10}}{\frac{1}{10} + \frac{7}{15}} = \frac{3}{17}$$

Index

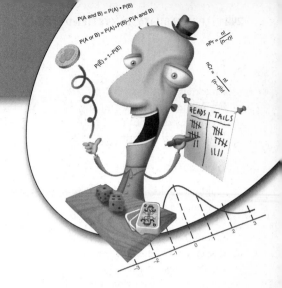